Additive Manufacturing
Fundamentals and Advancements

Additive Manufacturing
Fundamentals and Advancements

Manu Srivastava, Sandeep Rathee,
Sachin Maheshwari and T. K. Kundra

CRC Press
Taylor & Francis Group
Boca Raton London New York

CRC Press is an imprint of the
Taylor & Francis Group, an **informa** business

CRC Press
Taylor & Francis Group
6000 Broken Sound Parkway NW, Suite 300
Boca Raton, FL 33487–2742

First issued in paperback 2020

ISBN-13: 978-1-138-48545-7 (hbk)
ISBN-13: 978-0-367-77654-1 (pbk)

Library of Congress Cataloging-in-Publication Data

Names: Srivastava, Manu, author. | Rathee, Sandeep, author. | Maheshwari, Sachin, author. | Kundra, T. K., author.
Title: Additive manufacturing : fundamentals and advancements / by Manu Srivastava, Sandeep Rathee, Sachin Maheshwari, and T.K. Kundra.
Description: Boca Raton, FL : CRC Press/Taylor & Francis Group, 2019. | Includes bibliographical references. | Summary: "There is a growing need for manufacturing optimization all over the world, and the immense market of Additive Manufacturing (AM) technologies proves dictates a need for a book that will provide knowledge of the various aspects of AM for anyone interested in learning about this fast growing topic. This book will disseminate knowledge of AM amongst scholars at graduate level, post graduate level, doctoral level, as well as industry personnel. The objective is to offer a state-of-the-art book which covers all aspects of AM and incorporates all information regarding trends, history, principles, limitations and advancements in a one-stop resource"– Provided by publisher.
Identifiers: LCCN 2019020277 | ISBN 9781138485457 (hardback : acid-free paper) | ISBN 9781351049382 (ebook)
Subjects: LCSH: Three-dimensional printing. | Manufacturing processes.
Classification: LCC TS171.95 .S73 2019 | DDC 621.9/88–dc23
LC record available at https://lccn.loc.gov/2019020277

Visit the Taylor & Francis Web site at
www.taylorandfrancis.com

and the CRC Press Web site at
www.crcpress.com

Contents

Section C Material, Design and Related Aspects of
Additive Manufacturing Processes

Section D Trends, Advancements,
Applications and Conclusion

List of Figures

List of Tables

List of Abbreviations

.stl	Stereolithography (file format)
2D	Two-dimensional
2DP	2D printing
2PP	Two-photon polymerization
3D	Three-dimensional
3DP	3D printing
3F	Form, fit and function
3SP	Scan, spin and selectively photocure technology
ABS	Acrylonitrile butadiene styrene
AFS	Additive friction stir
AM	Additive manufacturing
ASTM	American Society for Testing and Materials
BJ	Binder jetting
Blisk	Bladed disk
BPM	Ballistic particle manufacturing
BT	Build time
CAD	Computer aided design
CAE	Computer aided engineering
CAM	Computer aided manufacturing
CAM-LEM	Computer aided manufacturing of laminated engineering materials
CAPP	Computer aided process planning
CIJ	Continuous stream inkjet
CJP	ColorJet Printing
CLI	Common layer interface
CLIP	Continuous liquid interface production
CLS	Conformal lattice
CMTs	Conventional manufacturing techniques
CNC	Computerized numerical control
CSAM	Cold spray-based additive manufacturing
CS	Cold spraying
DED	Directed energy deposition
DFA/M	Design for assembly and manufacturing
DFAM	Design for additive manufacturing
DFM	Design for manufacturing
DLD	Direct laser deposition
DLP-SLA	Digital light processing SLA process
DMD	Digital micrometer device
DMLS	Direct metal laser sintering
DOD	Drop-on-demand

EB-DED, EBF3	Electron beam-based DED process
EBM	Electron beam melting
E$_s$	Total free energy
FASW	Friction assisted seam welding
FBAM	Friction-based additive manufacturing
FD	Friction deposition
FDM	Fused deposition modelling
FFF	Fused filament fabrication machine
FGM	Functionally graded material
FS	Friction surfacing
FSAM	Friction stir additive manufacturing
FSLW	Friction stir lap welding
FSW/P	Friction stir welding/processing
GM	Generative manufacturing
HAM	Hybrid additive manufacturing
HPGL	Hewlett-Packard Graphics Language
HP	Highly-filled polymeric material
IGES	Initial Graphical Exchange Specification
IJP	Inkjet printing
ISO	International Organization for Standardization
LAM	Laser additive manufacturing
LB-DED	Laser-based directed energy deposition
LCD-SLA	Liquid crystal display stereolithography
LEAF	Layer Exchange Ascii Format
LENS	Laser Engineered Net Shaping
LFB	Liquid fusion binding
LFW	Linear friction welding
LIFT	Laser-induced forward transfer
LMD	Laser metal deposition
LMI	Layer manufacturing interface
LMJ	Liquid metal jetting
LOM	Laminated object manufacturing
LS	Laser sintering
MAM	Metal-based or metal additive manufacturing
MELD	Another name for AFS
MFI	Melt flow index
MJ	Material jetting
MJM	Multi-jet modelling
MJP	Multi-jet printing
MMCs	Metal matrix composites
M.T.	Melting temperature
MV	Model material volume
NASA	National Aeronautics and Space Administration
NIST	National Institute of Standards and Technology
NMJ	Nano Metal Jetting©
NPJ	NanoParticle Jetting

PA	Polyamide
PBF	Powder bed fusion
PBIH	Powder bed/inkjet head 3D printing
PC	Production cost
PCL	Polycaprolactone
PDM	Plasma deposition
PH	Print head
PI	Photoinitiator
PMC	Polymer-based composite
PPP	Plaster-based 3D Printing
PSL	Plastic sheet lamination
PUR	Polyurethane
RFW	Rotary friction welding
RP	Rapid prototyping
RP/T/M	Rapid prototyping/tooling/manufacturing
RPI	Rapid prototyping interface
S_a	Total surface area of particle
$SA_{bed}/VR_{particle}$	Surface area bed/volume ratio of particle
SASAM	Standardization of additive manufacturing
SDL	Selective deposition lamination
SGC	Solid ground curing
SLA	Stereolithography
SLA	Stereolithography apparatus
SLC	3D System SLice Contour
SLM	Selective laser melting
SLMR	Selective laser powder remelting
SLPs	Sheet lamination processes
SLS	Selective laser sintering
SM	Support material
SSS	Solid state sintering
STEP	STandard for Exchange of Product model data
STL	STereoLithography
Super IH SLA	Super integrated hardened polymer stereolithography
TCE	Time compression engineering
TE	Tissue engineering
TPP	Two photon photopolymerization
$T_{Processing}$	Processing temperature
UAM	Ultrasonic additive manufacturing
UC	Ultrasonic consolidation
UV	Ultraviolet
USW	Ultrasonic welding
VP	Vat photopolymerization
VP	Virtual prototyping
VRML	Virtual reality modelling language
γ_s	Surface energy per unit surface area

Preface

Rapid ongoing technical and technological manufacturing advances have necessitated the enhancement of efficiencies and efficacies at each level. The advent of computer aided design (CAD) and computer aided manufacturing (CAM) is decades old now and has gone a long way in compacting process cycles and enhancing process abilities as well as flexibilities. There is still a restraint on the degree of integration of these two core technologies. Computer numerical controlled (CNC) machines have successfully integrated them to an appreciable extent. However, despite numerous advantages offered, a complete amalgamation of CAD and CAM is not seen even with CNC technologies, owing to the substantial human intervention required for path planning, process planning and machine set-up. This gave birth to rapid prototyping (RP) techniques where complete integration of CAD and CAM is evident to fabricate parts with almost no restraint on complexity. RP technologies have witnessed immense growth over the past three decades and are now being used for full-fledged manufacturing and tooling applications. RP technologies are today better referred to as "additive manufacturing" or "AM techniques," though the term "RP" is still not obsolete. Additive manufacturing is a terminology for techniques based upon generative principles capable of producing layered artefacts with minimal intricacy and time constraints.

The quantum of research being accomplished in the field of AM and its commercial success is reflected in the huge growth of number of patents being filed, advent of newer AM techniques, published research on AM and its hybrid techniques, as also the newer applications being reported in almost each field. Extremely fast paced developments combined with accelerated obsolescence in this field makes it extremely challenging to present all the latest developments. However, exchange of ideas is a necessary part of any learning process. In this regard, everyone should be morally responsible to disseminate knowledge gained with the research and academic communities. This book is one such effort by the authors' and is aimed at dissemination of information as well as experience in the area of additive manufacturing.

The authors of this book started working together as a research team, and later they developed a keen interest in the area of additive manufacturing owing to the effectiveness of these techniques. The primary inspiration of the authors to undertake the responsibility of this book is the need to present a one-stop destination that can cater to the needs of academics, industries and the researcher community. Language is kept at the simplest technical form to render usefulness and interest to novice additive manufacturing professionals. This book is an attempt to provide an exhaustive and extensive compilation to enrich the reader's knowledge about AM processes. The authors have tried their best to provide exhaustive quality information by including basic aspects and major advances, as well as trends in AM. Sufficient literature has been reviewed and the knowledge of the authors based upon their several years of work experience has been utilized to write this book.

All the mandatory aspects of additive manufacturing have been discussed in detail in the present work. This book is divided into four sections and eighteen chapters.

Section A presents general details of AM processes and consists of five chapters.

Chapter 1 introduces the readers to the basics of AM, especially with respect to types of AM technologies, nomenclature of AM machines, advantages and limitations of AM technology, direct as well as indirect prototyping, tooling and manufacturing; it finally concludes with a summary of discussion.

Chapter 2 presents a comparison of the AM technology with conventional manufacturing processes with special discussion of CNC machining, deformation processes and primary shaping processes. It also presents the relative merits and limitations of AM technology over conventional manufacturing techniques and finally summarizes the discussion towards the end of the chapter.

Chapter 3 presents a detailed overview of the classification of AM processes on various bases such as physical state of raw material, processing techniques, underlying technology, fabrication technique, energy source, raw materials being used and material delivery system. Many commercially available AM processes are discussed along various parameters. The chapter concludes with a summary.

Chapter 4 gives a detailed account of the evolution of AM technology. It presents a detailed timeline of the AM processes and concludes with a summary.

Chapter 5 introduces the generalized AM process chain. It discusses the role of AM as a fundamental time compression engineering (TCE) element. It proceeds to discuss the AM data and information flow. It then presents details of the AM process chain including its eight steps; basic variations between various AM modellers; and a few maintenance and material handling issues. It concludes with a summary.

Section B presents process specific details of various AM processes and consists of seven chapters.

Chapter 6 details AM processes using vat photopolymerization in terms of various material related aspects, including precursors, photoinitiators, absorbers, filled resins, additives and post-processing; details of the photopolymerization process; process modelling aspects; variants and classification of the vat photopolymerization process including free and constrained surface approach, Laser-SLA, digital light processing SLA process (DLP-SLA), liquid crystal display stereolithography; and the advantages and limitations of vat photopolymerization processes. The chapter concludes the discussion with a summary.

Chapter 7 presents details of AM techniques utilizing powder bed fusion (PBF) processes including materials; powder fusion mechanism; process parameters and modelling; powder handling; powder fusion techniques including solid state sintering, chemical sintering, complete melting, liquid phase sintering/partial melting, indirect processing, pattern method and direct sintering; powder bed fusion process variants including low temperature laser-based processing, metal and ceramic laser-based systems, electron beam melting (EBM) and line- and layer-wise systems; and strengths and weaknesses of PBF-based AM techniques. The chapter concludes the discussion with a summary.

Chapter 8 gives details of AM processes utilizing extrusion-based systems including basic principles of extrusion-based processes; fused deposition modelling (FDM) including its performance measures and FDM limitations; bio-extrusion; contour crafting; non-planar systems; FDM of ceramics; RepRap FDM systems; and applications. It concludes the discussion with a summary.

Chapter 9 covers AM processes utilizing material jetting in terms of basic underlying principle; multi-jet printing; droplet formation techniques including continuous stream and drop-on-demand inkjet technology; materials for material jetting; advantages and limitations of material jetting; applications; and design and quality aspects. It concludes the discussion with a summary.

Chapter 10 covers AM processes utilizing binder jetting with respect to various aspects including process description; raw materials; design and quality aspects of binder jetting; advantages and limitations of binder jetting; and applications. It concludes the discussion with a summary.

Chapter 11 deals with AM processes utilizing sheet lamination processes in terms of its variants; laminated object manufacturing including its process description, materials, process variants, advantages, limitations and applications; ultrasonic consolidation including its basic principles, advantages, limitations and applications; and design and quality aspects. The chapter concludes the discussion with a summary.

Chapter 12 deals with AM processes utilizing directed energy deposition (DED) processes and covers the general DED process description; laser-based directed energy deposition techniques including direct laser deposition, laser-based DED techniques for 2D geometries and laser-based DED techniques for 3D geometries; applications of laser based DED (LB-DED) including laser-assisted repair, laser cladding and electron beam based DED processes; and advantages, limitations and applications of the DED process. It concludes the discussion with a summary.

Section C covers design and quality aspects of AM processes and consists of two chapters.

Chapter 13 presents a detailed discussion of the various materials for additive manufacturing; their forms/state including polymers, metals and their alloys, ceramic materials, composite materials, etc.; material binding mechanisms in AM including binding using secondary phase assistance, binding using chemical induction, binding using solid state sintering and binding using liquid fusion; and defects in AM parts including balling phenomena, porosity defect, cracks, distortion, inferior surface finish, etc. The chapter concludes the discussion with a summary.

Chapter 14 outlines the AM design and strategies and covers topics including design for additive manufacturing (DFAM), AM design tools, design considerations and DFAM system details, and concludes the discussion with a summary.

Section D covers solid state hybrid AM techniques, advancements, applications and conclusion, and has four chapters.

Chapter 15 covers solid state hybrid AM techniques including ultrasonic additive manufacturing, its working principle, benefits and applications; additive manufacturing using cold spraying including its working principle, advantages, limitations, applications and challenges; friction-based additive manufacturing including friction deposition-based, friction surfacing-based, linear friction welding-based, rotary friction welding-based, friction stir additive manufacturing, friction assisted seam welding-based additive manufacturing, additive friction stir; and conclusion and future scope of hybrid AM techniques. The chapter concludes the discussion with a summary.

Chapter 16 outlines AM applications, particularly with respect to application of AM parts as visualization tools, aerospace, automotive, medical, construction industry and retail applications. It concludes the discussion with a summary.

Chapter 17 presents a detailed discussion on the impact and forecasting of additive manufacturing including the influence of AM upon health and well-being, environment, supply chain management, health and occupational hazards, repair; economic characteristics of AM; sustainability of AM; and future of AM. It concludes the discussion with a summary.

Chapter 18 presents an overall conclusive summary to this book and also a detailed discussion on the future trends of AM. Many unexplored domain aspects have been identified and are highlighted in this chapter.

The authors sincerely wish that this book holds value for university students as a textbook, researchers, and academicians who plan to pursue a research career in the field of additive manufacturing. The authors will genuinely welcome and appreciate any queries, advice and observations by readers to add value to further editions of this book. The author group also genuinely prays to the Almighty that this book serves its purpose of benefiting students, academicians and the research community. We genuinely hope that readers can apply knowledge of the information presented in this book to promote research and development in the field of AM.

Summary

This book carries a detailed account of various fundamental and advanced aspects of additive manufacturing (AM). The four discrete sections of this book cover the fundamental concepts; process details; design, material and software aspects and advancements, applications and trends of additive manufacturing. The author team has tried their best to provide quality information based upon their teaching experience, extensive literature, research and market surveys. The author team have also recognized many unexplored additive manufacturing domains in their present work.

Features

1. Systematic coverage of various aspects of AM within four distinct sections.
2. Detailed explanation of various AM techniques on the basis of American Society for Testing and Materials (ASTM) guidelines using seven chapters each dedicated to an individual technique.
3. Discussion of numerous AM applications with suitable illustrations.
4. Detailed coverage of recent trends in the field of AM.
5. Engineering materials utilized as raw materials in AM are covered in detail.
6. Comparison of AM techniques with different traditional manufacturing methods is undertaken for clarity.
7. Review of relevant quality literature has been coupled with practical knowledge of the authors in finalizing and writing the contents of this book.

Acknowledgments

Words fail us when we try to express our affection and gratitude for the editor Ms. Cindy Carelli for agreeing to publish this work and for her immense support, cooperation as well as coordination. She is inarguably a thorough professional who can add value to any project she agrees to be a part of. The authors are in complete awe of her knowledge and efficacy but what is most important about her is the beautiful and caring golden heart which she possesses. Her ability to understand the value of work and life is something that makes her a completely unique human being. The authors thank CRC Press for their support towards this initiative of bringing forth this book.

Dr. Manu Srivastava wishes to thank her Vice President Dr. Amit Bhalla, Vice Chancellor Dr. Sanjay Srivastava, Director General Dr. N. C. Wadhwa, Pro-Vice Chancellor and Dean FET Dr. M. K. Soni for placing their immense trust in her capabilities. She also thanks her university officials and faculty members for their inspiration and support. The author wishes to convey her heartfelt thanks to her supervisors Dr. T. K. Kundra and Dr. Sachin Maheshwari. A heartfelt thanks is also due to her mentor in the field of AM Dr. Pulak Mohan Pandey. Dr. Pandey is one example of rare combination of modesty, a beautiful heart and brilliant intellectual capabilities. She takes this platform to acknowledge the support of her friends and family, specially her late parents, Yash Krishna Srivastava, Mr. Ram Krishna Yashaswi, Dr. Sandeep Rathee, Dr. T. K. Singh, Ms. Rachna Johar, Ms. Shilpi Chauhan, Ms. Vartika and Ms. Neelu for their constant support and unfaltering trust.

The author Dr. Sandeep Rathee wishes to thank his Vice Chancellor Lt. Gen. V. K. Sharma, AVSM (Retd.); Pro-Vice Chancellor Prof. (Dr.) M. P. Kaushik and Director Maj. Gen. S. C. Jain, V. S. M**(Retd.) for placing their immense trust in his capabilities. He thanks his friends Mr. Nasir Khan and Mr. Neeraj Saini. The author wishes to convey his heartfelt thanks to his supervisors Dr. Sachin Maheshwari and Dr. Arshad Noor Siddiquee. He also acknowledges the support of his family, specially his parents Mr. Raj Singh Rathee and Mrs. Krishna Rathee, Dr. Manu, Mr. Sombir Rathee, Mrs. Manju and Mrs. Anita for their constant support.

Dr. Sachin Maheshwari wishes to thank his family, specially his mother, his wife Mrs. Monica Maheshwari and his children. He also acknowledges their support towards the completion of this book. A heartfelt thanks is also due to all the academic and institutional colleagues for their necessary support.

Dr. T. K. Kundra wishes to thank his family, specially his late parents, his wife and his children. He also acknowledges their support towards the completion of this book. A heartfelt thanks is also due to all the academic and institutional colleagues for their necessary support.

Finally, the authors devote and dedicate this work to the divine creator. They thank the Almighty for giving them the strength to bring these thoughts and understanding of concepts into physical form. The authors pray that this work may be of enough technical competence to enlighten the readers about each relevant aspect of additive manufacturing techniques.

Dr. Manu Srivastava
Dr. Sandeep Rathee
Dr. Sachin Maheshwari
Dr. T. K. Kundra

About the Authors

Dr Manu Srivastava is presently serving Manav Rachna International Institute of Research and Studies, Faridabad, India as a Professor in the department of mechanical engineering. She has completed her PhD in the field of additive manufacturing from the department of Manufacturing Processes and Automation Engineering in Netaji Subhas Institute of Technology, New Delhi. Her field of research is additive manufacturing, friction-based AM, friction stir processing, automation, manufacturing practices and optimization techniques. She has more 50 publications in international journals of repute and refereed international conferences. She is the author of *Friction Based Additive Manufacturing Technologies: Principles for Building in Solid State, Benefits, Limitations, and Applications* (CRC Press, 2018). She has a total experience of around 14 years in teaching. She has won several proficiency awards during the course of her career including merit awards, best teacher awards, etc. The following courses are taught by her at graduate and postgraduate level: Robotics, Manufacturing Technology, Advanced Manufacturing Processes, Material Science, CAM, Operations Research, Optimization Techniques, Engineering Mechanics, Computer Graphics, etc. She is a life member of the Additive Manufacturing Society of India (AMSI), Vignana Bharti (VIBHA), The Institution of Engineers (IEI India), Indian Society for Technical Education (ISTE), Indian Society of Theoretical and Applied Mechanics (ISTAM) and Indian Institute of Forging (IIF).

Dr. Sandeep Rathee is currently working as an Assistant Professor at the Amity School of Engineering and Technology, Amity University Madhya Pradesh, Gwalior, India. He has been awarded a PhD degree from the University of Delhi. His PhD work is in the field of friction stir welding and processing from Department of Manufacturing Processes and Automation Engineering, Netaji Subhas Institute of Technology, New Delhi. His field of research mainly includes friction stir welding/processing, additive manufacturing, advanced manufacturing processes and optimization. He has authored/co-authored over 40 publications in reputed international journals and refereed conferences. He is the author of *Friction Based Additive Manufacturing Technologies: Principles for Building in Solid State, Benefits, Limitations, and Applications* (CRC Press, 2018). He has a total teaching experience of around eight years. He is a life member of the Additive Manufacturing Society of India (AMSI) and Vignana Bharti (VIBHA).

Dr. Sachin Maheshwari is currently serving at Netaji Subhas Institute of Technology as a professor in the division of Manufacturing Processes and Automation Engineering and as Dean, Faculty of Technology, University of

Delhi. He has completed his PhD from the Indian Institute of Technology, Delhi in the field of welding and ME in Industrial Metallurgy from the Indian Institute of Technology, Roorkee. His areas of interest include variants of welding, advanced manufacturing especially additive manufacturing, optimization techniques and unconventional manufacturing techniques. He has over 100 research papers in international journals and refereed conferences. He is the author of *Friction Based Additive Manufacturing Technologies: Principles for Building in Solid State, Benefits, Limitations, and Applications* (CRC Press, 2018). He has guided several PhD theses; about ten PhD theses have been awarded and equal numbers are under progress. He has a total teaching and research experience of around 23 years and has taught a wide assortment of subjects during his teaching career. He has rich experience of working on the statutory authorities and experience of handling academic assessment and accreditation procedures. He is associated with many research, academic and professional societies in various capacities.

Dr. T. K. Kundra has served the Indian Institute of Technology Delhi as Professor and Head of Department of Mechanical Engineering and continues to serve the same world-renowned institution. His areas of interest are optimal mechanical system design including micro systems and computer integrated manufacturing systems including additive manufacturing. His experience of teaching, research and design is spread over 48 years and includes teaching at AIT Bangkok, Addis Ababa University and studies at Loughborough University, Imperial College, Ohio State University, TU Darmstadt. He has been consulted by several organizations such as Hero Motors, BHEL, Eicher, ONGC, DRDO and the British Council. He is a Chartered Engineer and Fellow of the Institution of Engineers. He is the author/co-author of more than 100 technical research papers and the co-author of two text books on Numerical Control and Computer Aided Manufacturing (*Numerical Control & Computer Aided Manufacturing*, Tata McGraw-Hill Education, 1987, and *Computer Aided Manufacturing*, Tata McGraw-Hill Education, 1993), and two books on Optimum Dynamic Design and friction-based additive manufacturing (*Optimum Dynamic Design, Allied Publishers Limited, 1997 and Friction Based Additive Manufacturing Technologies: Principles for Building in Solid State, Benefits, Limitations, and Applications*, CRC Press, 2018). Also, he has been awarded the honor of "Mechanical Engineer of Eminence." He has introduced/developed/taught a wide spectrum of subjects (around 40) in his teaching career at graduate/post graduate level including mechanical design and optimization, plant equipment design, CNC manufacturing. He is associated with many professional societies in different capacities.

Section A

General Details of Additive Manufacturing Processes

1

Introduction

1.1 Introduction

AM is an automated manufacturing technique for fabricating three-dimensional (3D) layered artefacts directly from computer aided designs without the need of any tool, jig or fixture. AM, along with subtractive techniques (milling/turning) and formative fabrication (casting/forging), form an integral part of today's manufacturing environment. During its early years, this technique was called three-dimensional printing (3DP), a name which retains popularity even today despite it being a misnomer. Terms like generative manufacturing (GM) and rapid manufacturing (RM) are very popular synonyms of AM used by various sectors. AM has come a long way over the past 30 years to metamorphize into its current stature. This development is consistent and progressive. Advanced AM technologies and modellers have been burgeoning, resulting in market diversity with respect to technological compatibility as well as competence.

This is the first chapter of Section A of this book and presents general details of AM processes. It aims to introduce the readers to the basics of AM, specially with respect to types of AM technologies, nomenclature of AM machines, advantages and limitations of AM technology, direct as well as indirect prototyping, and tooling and manufacturing. It concludes with a summary of discussion.

1.2 Types of AM Technologies

Various types of AM processes have been defined on different bases. Different AM personnel have presented various versions towards this objective. A summary of these is given below for detailed and quick reference:

1. Direct and indirect application AM processes based upon application levels.

2. Metallic and non-metallic AM processes based upon type of modeller/raw material used.

3. Solid, liquid, powder or gaseous AM processes based upon the physical state of raw material utilized.

4. One-dimensional, array of one-dimensional and multi-dimensional arrays based upon data transfer mechanism from stereolithography (STL) data format to the modeller.

5. Resin photopolymerization, material extrusion, directed energy based, building printing, sheet lamination and powder bed binding/fusion based upon the working principle or underlying technique.

6. Fusion based and solid state AM on the basis of the state of raw material during the process of component fabrication and the technique employed for addition of subsequent layers.

7. AM based on binder, laser, heat of friction, plasma, beam of electrons, etc. on the basis of energy source used during the process.

8. Plastic, ceramic, powder, resin, etc. AM on basis of raw material used.

9. Powder bed and powder feed AM machines on basis of material delivery system utilized.

10. VAT photopolymerization (VP), material extrusion (ME), binder jetting (BJ), material jetting (MJ), sheet lamination (SL), powder bed fusion (PBF) and directed energy deposition (DED) as per ASTM F42 guidelines.

Apart from this, other types of AM processes are also coming up owing to exponentially increasing technical growth as well as obsolescent rates in the field of AM.

1.3 Nomenclature of AM Machines

A wide variety of AM machines are commercially available in the market today. Application level and type of AM process constitute two major bases for their nomenclature. Another common classification is based upon infrastructure, operating requirements and mean price. The following noteworthy points are presented for reference of readers:

1. If a machine is capable of fabricating parts, it is termed a "fabricator."

2. If its capability is limited to fabricating prototypes, then it is called a "prototyper."

3. "Printer/3D printer" is generally used for all AM machines. It is prefixed with application genre, i.e. personal/professional.

4. In general, all AM machines can be put into three main categories which are: fabber, office and shop-floor AM machines.

5. A fabber is a small, modest and low-cost modeller. If it is used personally/privately by a single/multiple person from home/co-working domain, then it is called personal fabber.

6. Its application is mainly confined to producing prototypes, solid images or concept models.

7. If an AM machine is used in an office, then it is called an "office AM machine" and is characterized by minimal noise, odor, etc. Refilling of the model/support material is undertaken by office staff and is typically in cartridge form. Simple part handling, easy operation, minimal maintenance, easy waste disposal, uncomplicated post-processing, etc. are some key features of this class of machines.

8. Its application is mainly for prototyping (primary or functional or masters for secondary RP processes).

9. Sophisticated workspace, industrial ambience, complex logistics, skilled labor characterizes an industrial shop-floor or simply shop-floor machine. Larger raw material range, enhanced productivity and outputs, complex post-processing requirements are key features of shop-floor machines. Economics of production need to be analyzed in depth in these machines.

10. Their use is spread over manufacturing, prototyping, as well as tooling.

11. The term "printer/modeller/fabricator/3D printer" is generally used for these different AM machines as a general convention.

12. The price of fabbers is less than that of office AM modellers, which in turn is less than that of shop-floor AM machines.

13. Operating skills required are minimal in case of fabbers. Anyone professionally qualified in CAD software can work with office AM machines. However, shop-floor machines require proper trained personnel.

1.4 Prototyping, Tooling and Manufacturing

This section provides a basic overview of the correct terminology to be utilized in the three main AM based domains, i.e. prototyping, tooling and manufacturing. A common myth surrounding this field is that it is mostly thought of as process based field rather than application-based approach. It should be clearly understood here that various AM techniques can cater to a specific application. Thus, contrary to the popular belief that various processes need to understood as a first step, it needs to be remembered that the application needs to be completely understood as the first step in the decision regarding fabrication via the AM route. After this, the intricacies of the design for a particular part need consideration. It is only after this detailed analysis that one can conclude which particular AM technique needs to be adopted that can accommodate all the requisite part requirements. The above discussion clearly indicates that various broad areas of applications need to be clearly understood in a systematic manner. This, in turn, requires defining the application levels precisely. The levels defined here are still not standardized but are considered key indicators by some prestigious researchers in the field of AM.

1.4.1 Direct AM Processes

AM technology is divided into three main application levels which are prototyping, tooling and manufacturing.

1.4.1.1 Direct Prototyping

Prototyping encompasses AM applications based on fabricating any prototype, concept model, specimen or mock-up. It has two sub-levels which are:

1. Solid imaging and conceptual models that are applied for verifying concept and can be compared to meagre three-dimensional printouts of the design; they are also known as show-and-tell models. To verify intricate drawings, scaled concept models (also called data control models) are used. Colored models utilized to evaluate designs also fall under this category.

2. Functional prototypes are used to check and verify functions of a limited number of parts of final product during production decisions. An example of this is an adjustable air outlet grill for adjusting car nozzle to impart flexibility in responding to climatic changes during the early product development phase.

1.4.1.2 Direct Tooling

Tooling encompasses AM applications based on fabricating cores, cavities, inserts for tools, dies and molds, etc. For an AM part to be understood as final, it should have all necessary traits and functionalities specified during the product development phase. If the final part is a negative or inversion of data set, i.e. die/mold/gauge, then the technique is called "Rapid tooling" and can be further categorized into two types, i.e. direct or prototype tooling. It is confined to the tooling applications only and is generally based upon metal additive manufacturing (MAM). A key point to be made clear is that the complete tool is not obtained but only its important parts are created. The tool is obtained via conventional manufacturing route here, also by utilizing its parts like cavities, cores, etc. AM tooling facilitates the easy fabrication of inner cavities for features such as cooling channels beneath an outer surface. Such cooling, also called conformal cooling, enhances cooling design and mold cavity productivity manifolds.

1.4.1.2.1 Prototype Tooling

Making a mold is an expensive and time-consuming affair, especially for manufacturing in small batches and where rapid design changes need to be incorporated. In such cases, we use a temporary mold that possess the traits of functional prototypes and at the same time is suitable for direct tooling application. This gives rise to an intermediate application level which is frequently known as prototype or bridge tooling.

It needs to remembered that rapid tooling in itself is not an autonomous application level but denotes the integration of AM applications which can fabricate the die/mold or its insert.

1.4.1.3 Direct Manufacturing

Manufacturing encompasses applications based on obtaining a final part or end product that can either be used directly or suitably assembled. If the resulting part (as discussed above) is a positive, the AM technique becomes "direct manufacturing." A necessary checkpoint here to ensure aptness of design is to recheck it against chosen process details as specified during initial engineering design.

Most AM processes come under genre of direct AM since 3D CAD models are used to directly obtain physical parts. There are a few indirect AM processes also where a layered approach to obtain artefacts is not utilized. These are based upon obtaining an initial master sample via AM and then copying/cloning this sample. Examples of indirect AM techniques include silicon rubber casting-based techniques.

1.4.2 Indirect AM Processes

Any AM process can be used to obtain an almost exact facsimile of object from a CAD data set. These processes are, however, accompanied by some limitations which include:

1. These are process and in turn machine dependent.
2. There are many restrictions on AM materials.
3. Unlike conventional manufacturing, there is minimal reduction in cost with increased volumes of production.
4. They become increasingly expensive if multiple copies of the same component need to be made and so on.

One probable solution is to utilize AM parts as master models at the first stage and then copying/replicating/reproducing them at a second stage. The splitting of capabilities is the concept behind this technique. While stage 1 is a thoroughly additive process in nature, stage 2 does not follow the additive principle at all. Such a process is therefore called "indirect RP/indirect AM process" due to its indirect nature.

1.4.2.1 Indirect Prototyping

Indirect prototyping is utilized to improve AM part characteristics to suit specific customer demand. For example, if a specific colored part is required that is not compatible with the available AM facility, then we can make a rigid master model of any color and use it for further casting of the required component. Due to the need for detailed surface and fully loaded parts during copying, only a few AM techniques, such as polymer jetting, are used to create master models.

Indirect prototyping techniques rarely find utility in solid imaging or concept modelling owing to cost considerations. This technique is mostly employed for functional prototyping.

1.4.2.2 Indirect Tooling

Indirect tooling is also based upon similar copying procedures as explained above. Though the final part is not obtained, the tool forming the basis for producing small/medium-sized batches of the end part is obtained. Indirect tooling offers considerable saving in time and cost over conventional methods of tool fabrication. Unlike indirect prototyping, this tool is used for final components and can be utilized for a large number of components. Milling inserts can be utilized for improving a sharp edge.

An example of this is a mold for making wax patterns for lost wax casting obtained from an AM master by counter-casting it in polyurethane, PUR, backed by an aluminium box. After the AM master is removed, the mold is used to process the required amount of wax pattern. The higher rigidity of the PUR material in combination with the backed-up walls leads to a mold that delivers much more precise wax patterns than could be made by a soft silicon mold. In comparison to milled all-aluminium tools it is cheaper and has a much shorter lead time. This kind of mold can be used for a small series production of complex precision parts.

1.4.2.3 Indirect Manufacturing

Indirect manufacturing also finds its basis in the creation of AM master parts. The objective is to create final parts with properties equivalent to traditionally manufactured products.

1.5 Advantages of AM Processes

AM provides a unique ability to fabricate components with high variability and flexibility in geometrical features. It offers a path of fabricating some special components like light hollow contours or mold cavities with passages for internal cooling, etc. Great cost savings (more than 50% in general in aerospace/automobile industries) can be obtained by the use of the AM route for part fabrication as compared to the conventional methods of manufacturing. The time required to bring the component to market is greatly reduced by this route due to the enormous compaction of the design cycle in case of AM. Appreciable strength-to-weight ratio metallic parts can be fabricated since a high degree of freedom in design is permitted by the AM route. The quality of parts in terms of features and intricacy is highly improved. These are environmentally friendly manufacturing methods for two main reasons: (1) tremendous reduction in scrap and wastage which can be attributed to its mode of operation and (2) reduced noise and pollution allowing them to be easily employed in office environments rather than specially designed workshops. Apart from these, there are numerous advantages of AM processes, some of which are summarized below [1–6]:

- Since the components are fabricated in a layered fashion so there is no requirement for tools and fixtures.
- The intricacy of the component has neglible impact on the time and cost of the final AM product in contrast to the conventional manufacturing processes.
- Nesting/parallel processing of parts is possible by careful layout optimization.
- Tremendous reduction in the lead times for part fabrication results in considerable cost and time savings.
- Highly customized parts can be made easily by the AM route.
- Design and fabrication of functionally graded materials is very easy by the AM route and extremely economical as compared to conventional manufacturing techniques.
- Multiple set-ups for part fabrication are not required in almost all cases.
- Operator intervention work is greatly reduced to a supervisory level.
- AM processes are highly responsive to dynamic manufacturing environment.
- Set-up/machine preparation time in case of AM processes is appreciably less as compared to that in conventional manufacturing.

Some other important advantages of AM include:

- Noise free
- Can be operated from the comfort of home or office
- Offers an excellent and impressive spectrum of applications
- Can form process chains when suitably combined with other conventional/unconventional manufacturing processes
- Lesser time for products to reach markets for customer end use
- Reduced material wastage owing to non-occurrence of mistakes
- Lesser costs owing to appreciable manufacturing savings
- Improved qualities
- Parts with complex and intricate geometries can be obtained
- Tools, molds or punches not required.

1.6 Challenges of AM Processes

Despite remarkable progress in the domain of AM, a variety of aspects like production speeds, build scale economies, precision, quality, raw materials, communication interfaces, etc. need attention to fully explore the potential of AM. Apart from the benefits stated in the previous section, AM is faced with multiple issues that restrict its use in a few application areas. The AM challenges are summarized as:

1. Non-optimal build speeds
2. Relatively less accuracy
3. Decision regarding optimal part orientation
4. Restricted choice of raw materials and resulting material properties
5. Poor surface finish
6. Pre-processing and post-processing requirements
7. High system cost chiefly owing to limited buyers
8. Anisotropic behavior of AM fabricated parts
9. Occurrence of stair-stepping phenomenon
10. Need for optimal layer thickness selection
11. Need of support structures
12. Poor structural strength of parts fabricated via AM techniques.

1.7 Summary

This chapter covers the various introductory aspects of AM. It starts with a formal introduction, which is followed by a discussion of various types of AM machines. A detailed discussion of the various direct as well as indirect prototyping, tooling and manufacturing application levels of AM-based processes follows next. Finally, advantages and limitations of AM are listed before a final conclusion. A comprehensive comparison of AM with conventional and advanced machining and manufacturing techniques on various bases is presented in Chapter 2.

References

1. Pham, D. T., Dimov, S. S., *Rapid manufacturing: the technologies and applications of rapid prototyping and rapid tooling*, 2001, Springer.
2. Gebhardt, A., Understanding additive manufacturing, in *Understanding additive manufacturing*, 2011, Hanser, pp. I–IX.
3. Gibson, I., Rosen, D. W., Stucker, B., *Additive manufacturing technologies: rapid prototyping to direct digital manufacturing*, 2010, Springer, pp. 1–459.
4. Srivastava, M., Maheshwari, S., Kundra, T. K., Virtual modelling and simulation of functionally graded material component using FDM technique. *Materials Today: Proceedings*, 2015, 2(4): pp. 3471–3480.
5. Srivastava, M., Maheshwari, S., Kundra, T. K., Rathee, S., Yashaswi, R., Sharma, S. K., Virtual Design, modelling and analysis of functionally graded materials by fused deposition modelling. *Materials Today: Proceedings*, 2016, 3(10, Part B): pp. 3660–3665.
6. Srivastava, M., *Some studies on layout of generative manufacturing processes for functional components*, 2015, Delhi University.

2

Comparison of Additive Manufacturing with Conventional Manufacturing Processes

2.1 Introduction

Additive manufacturing (AM) techniques basically possess many distinct features as compared to conventional formative and subtractive fabrication processes. AM technology is a bottom-up fabrication approach in which a part is built layer by layer into its designed shape. AM as a layered manufacturing method offers freedom in design and manufacturing of complex geometrical shapes, hybrid and functionally graded structures, etc. with high precision that cannot be matched by traditional manufacturing processes [1]. Most conventional fabrication processes are decades old and have now gained a stature where they have accomplished symphony with science and engineering material. In most cases, conventional techniques successfully add much value to the part fabricated so that it can retain its suitability for end use in terms of purpose and life span. However, the implementation of customized properties related to material size, complexity, shape and functionality is still a huge challenge for most conventional processes [1].

Conventional manufacturing techniques (CMTs) require a large work force or machinery and a high degree of supply chain management. On the other hand, AM techniques generally works on CAD software to print the final product with a wide range of materials and with minimum need for supply chain management. Also, AM does not need much in the form of costly molds, tools, jigs or fixtures, which therefore results in appreciable cost saving. In addition, the waste material in AM is comparatively less (around 40% reduced material wastage in metallic uses) than subtractive manufacturing processes. Furthermore, the recycling of around 95–98% of waste material is possible in AM [2]. According to a report by Airbus Group Innovations and their partner groups (UK), with the use of AM, the usage of raw material can be reduced by up to 75% for manufacturing specific products as compared to traditional manufacturing means [3]. General Electric (GE) reported a reduction of about 25% in cost and manufacturing time with the use of AM without compromising on performance [4].

This is the second chapter of Section A of this book, which presents general details of AM processes. In the previous chapter, all the basic details of AM processes are covered. This chapter aims to enable its readers to draw a clear comparison of the AM technology with conventional manufacturing processes, with special discussion on CNC machining, deformation processes, primary shaping processes, etc. It also presents the relative merits and limitations of AM technology over conventional manufacturing techniques and finally summarizes the discussion towards the end of this chapter.

2.2 Comparison between AM and Conventional Manufacturing

AM techniques generally imply a cluster of machines based on manufacturing processes where component fabrication process is achieved directly from CAD data using the principle of layer-by-layer additive joining. CMTs involve conventional manufacturing steps such as design, tooling and fixturing, fabrication, etc. to develop the product. A basic comparison between flow process of AM and CMTs is presented in Figure 2.1. This section briefly compares AM and CMTs, mainly subtractive manufacturing, forming and plastic deformation.

2.2.1 Comparison between AM and CNC Machining

A variety of subtractive manufacturing machines are available including routing, milling, turning, drilling, lathe and so on. Similarly, over hundreds of AM modellers utilizing varied techniques are commercially available today. In the subsequent paragraph, a brief comparison between AM and subtractive manufacturing, especially computer numerical controlled (CNC) machining is presented.

There are a few similarities between AM and CNC machining, some of which are presented below.

- Both can meet early concept modelling and prototyping requirements without delay and at nominal costs.
- Both can be used for a wide range of products from varied raw materials.
- Both processes possess almost identical workflows at broader levels, encompassing aspects like initial CAD drawing using 3D CAD software, machine instruction and machine preparation followed by actual fabrication of parts. However, it is executed in a different manner in the two technologies.

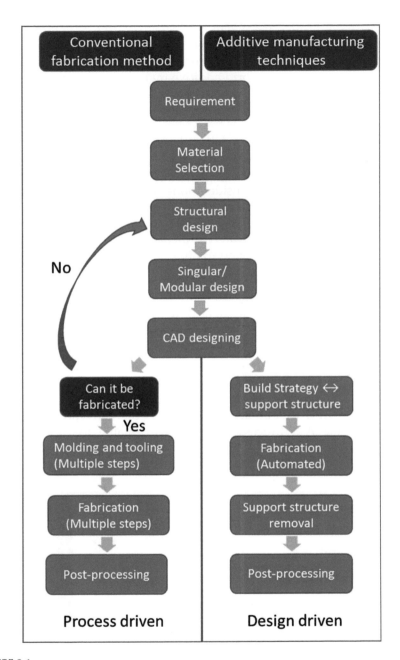

FIGURE 2.1
Basic Difference between Flow Processes in Conventional Manufacturing and Additive Manufacturing Processes. (*Source:* [5]. Reproduced with permission from Goh et al., Additive manufacturing in unmanned aerial vehicles (UAVs): challenges and potential. *Aerospace Science and Technology*, 63: pp. 140–151, Copyright 2017, Elsevier Masson SAS, all rights reserved.)

Despite these similarities, the two technologies are quite different from each other. The main point of distinction between these two technologies lies in the fact that while AM is an additive process, or may be additive and subtractive process (in the case of hybrid additive manufacturing processes), CNC machining is a subtractive process based upon conventional manufacturing. Some major points of difference between these two techniques are summarised in Table 2.1.

The schematic arrangement of AM and CNC machining process chains is shown in Figure 2.2 and an illustrative example of fabricating a specific part and the amount of scrap in these two techniques is presented in Figure 2.3.

Thus, there are manifold considerations that need to be weighed judiciously before finalizing a particular kind of manufacturing strategy. These mainly include: part intricacy, customization need, production time, dimensional accuracy, flexibility, overall cost incurred, part size, material availability, speed, tolerance, quality and so on.

2.2.2 Comparison between AM and Deformation Processes

Deformation processes generally deform the material plastically and transform it into a desired shape and size with defined properties. These processes can be classified into bulk forming and sheet forming processes. Bulk deformation processes mainly include extrusion, rolling, forging, etc., while sheet forming processes mainly include drawing, stretching, contouring, etc. [7]. Depending on the aspects of geometrical complexity and fabricated part properties, AM techniques are different from conventional deformation ones. While AM processes are highly suitable for manufacturing

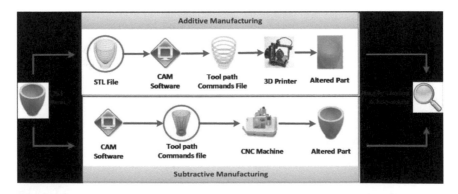

FIGURE 2.2
Process Chain for Additive and Subtractive (CNC Machining) Processes. (*Source:* [6]. Pan et al., Taxonomies for reasoning about cyber-physical attacks in IoT-based manufacturing systems. *International Journal of Interactive Multimedia and Artificial Intelligence*, 2017, 4(3), under Creative Commons Licence.)

TABLE 2.1

Differences Between AM and CNC Machining

Additive manufacturing	Computer numeric controlled machining
Based on subsequent addition of material	Based on subsequent removal of material
Based upon additive principle which is entirely different from conventional manufacturing	Based upon conventional manufacturing but with difference in control mechanism
Does not require tooling systems (tool, jig, fixture, etc.)	Requires sophisticated tooling system (tool, jig, fixture, etc.)
Parts are parallelly processed/nested to save time	Nesting is not possible
Number of setups is usually one thus reducing time lags	Multiple setups are generally required before obtaining final product leading to time lags
Comparatively slower in-process construction	Much faster process speed
File preparation is an easy task and does not require any specialized machining experience or programming skills	File preparation is a complicated task and requires specialized training and programming skill set
Import of CAD data into machine follows a simple .stl conversion route followed by some process-specific outputs and needs minimal manual inputs	Import of CAD data into CAM program is an intricate process and requires a lot of manual inputs
Modeller preparation is a simple process	CNC station preparation is a relatively time-consuming process
During real-time manufacturing, negligible human intervention is required	A lot of human intervention in form of reorientation and refixturing is necessary, specially when job shape is such that it is not fully accessible to operator
Material density and part height greatly affect build time	Feature count as well as intricacy specially affect build time
Most AM processes are accompanied by one or other kind of post-processing operation	Post-processing is rarely mandatory in case of CNC machining
Very low material wastage since material is used with great precision in only requisite quantities	A lot of material is wasted and clean-up is messy
Constraint on size of component to be fabricated owing to build volume limitations	No constraint on size of component to be fabricated
No constraint on complexity of component to be fabricated	Constraint on complexity of component to be fabricated owing to complicated programming and tooling requirements
Relatively poor surface with lesser transverse strength of components mainly due to stair stepping effect	High-quality parts with excellent surface finish can be obtained
Material choice is relatively constrained	Wide variety of materials are worked upon
AM is better for customized production	CNC is better for mass production owing to less time and overall cost
AM parts tend to undergo warping and distortion and also occurrence of layered contours	CNC parts have better polished looks and are rarely distorted during machining

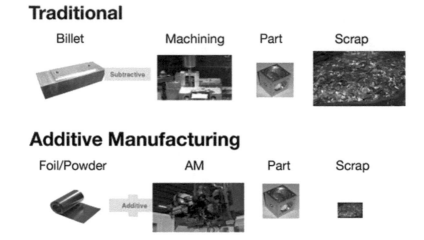

FIGURE 2.3
Illustration of Fabricating a Specific Part and the Amount of Scrap in AM and CNC Techniques.
(*Source:* Courtesy of Fabrisonic.)

complex geometrical components possessing anisotropy and involving higher production times, deformation processes are suitable for comparatively moderate complex structures with wrought microstructures with reduced production time.

2.2.3 Comparison between AM and Primary or Shaping Processes

Amongst the primary or shaping processes, casting and sintering are considered here. These two kinds of manufacturing, i.e. AM and shaping processes, share the similarity of consumption of raw material required for manufacturing a product. In AM processes, there is negligible wastage of materials and in casting and sintering processes, also, the raw material is shaped into the desired shape with little post-fabrication machining. The difference in these two kinds of manufacturing domains lies in their working processes, properties of fabricated parts, cost and complexity suitability. These aspects depend on the type of particular processes in respective domains.

2.3 Pros and Cons of AM with Respect to Conventional Manufacturing

In comparison to traditional manufacturing processes, AM offers numerous advantages, some of which are discussed here [8].

2.3.1 Part Flexibility
In AM techniques, there is minimal need for post-processing and also no constraints on tooling, so that complex geometrical shapes can be successfully obtained. In other words, in AM, part functionality is not sacrificed on the grounds of ease of manufacturing as is done in conventional manufacturing. Thus, for complex geometrical shapes, AM offers more suitability as compared to CMTs. Also, the cost of products with increased complexity remains the same in AM techniques which sharply increases in CMTs, as shown in Figure 2.4. Further, functional properties can be easily obtained in parts fabricated via AM techniques, which altogether opens up newer avenues for novel designs and innovations in manufacturing.

2.3.2 Waste Prevention
AM techniques utilize a layer-by-layer additive process for developing parts/components thereby reducing material wastage, which is a major issue in subtractive manufacturing processes where a large amount of material is removed during machining.

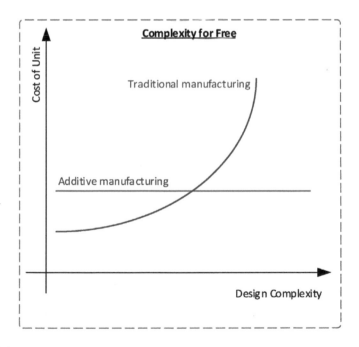

FIGURE 2.4
AM Complexity for Free. (*Source:* [9, 10]. Reprinted from Busachi et al. A review of Additive Manufacturing technology and Cost Estimation techniques for the defence sector, *CIRP Journal of Manufacturing Science and Technology*, 19: pp. 117–128, Copyright 2017, with permission from Elsevier. Hopkinson and Dicknes, Analysis of rapid manufacturing—using layer manufacturing processes for production. *Proceedings of the Institution of Mechanical Engineers, Part C: Journal of Mechanical Engineering Science*, 2003, 217(1): pp. 31–39.)

2.3.3 Production Flexibility
AM techniques require less costly set-up and do not require tooling and fixturing. Owing to this, AM techniques are economical and are suitable for small batch production. The properties of a fabricated part depend on the process and raw materials instead of operator skills. In addition, issues of production bottlenecks and line balancing are eliminated as AM techniques can produce complex shapes in a single piece.

2.3.4 Process Running Cost
Most of the subtractive manufacturing techniques need substantial labor, money and time for preparing fixtures, tools, molds, machine set-up, etc. owing to which the running cost per product is high. On the other hand, in the absence of tooling requirements and presence of shorter product development cycle, the running cost per product in AM techniques is small.

2.3.5 Probability of Change
New product/design development via traditional manufacturing techniques can be expensive, owing to many trial experiments and consequent time consumption. In contrast, AM offers freedom to design at negligible cost. Once a prototype is designed/developed by varying the CAD data, actual production can be performed. During AM, no extra cost in the form of retooling is involved.

Despite several benefits of AM techniques over traditional manufacturing processes, they present some challenges which are discussed below.

2.3.6 Start-up Investment
Owing to the high cost of AM modellers and the complex set-ups required during installations, the start-up investment is higher in the case of AM machines. However, conventional machines can be bought and set up at much lower prices than their AM counterparts.

2.3.7 Mass Production
During the production process, AM is less efficient in developing products at large scale as compared to traditional manufacturing. This can be seen from Figure 2.5 which shows that the initial product cost in AM is less and almost remains the same with an increase in production scale. This is mainly due to the slow deposition rates and built capacity constraints in AM [11]. On the other hand, the initial product cost in conventional processes is very high but reduces with an increase in scale of production.

2.3.8 Raw Material
With the recent innovations in AM processes such as selective laser sintering (SLS), laser beam machining (LBM), etc. it becomes quite possible to

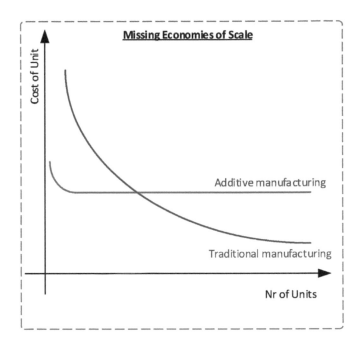

FIGURE 2.5
AM Economies of Scale. (*Source:* [9]. Reprinted from Busachi et al. A review of additive man-ufacturing technology and cost estimation techniques for the defence sector, *CIRP Journal of Manufacturing Science and Technology*, 19: pp. 117–128, Copyright 2017, with permission from Elsevier.)

process metallic components via 3D printing. However, some metal alloys still cannot be worked upon, or their output characteristics are poor. Thus, the range of raw materials in AM techniques is relatively smaller.

In comparison to AM, traditional manufacturing processes are well established and can process almost all types of materials.

2.4 Summary

Based upon different criteria, a comparison between AM and traditional manufacturing techniques is summarized in Table 2.2.

In this chapter, a detailed comparison of AM technology with conventional manufacturing processes is drawn. The pros and cons of AM technology over conventional processes is highlighted and brought out with suitable examples. The next chapter presents a detailed classification of AM processes on various bases. It also introduces readers to various commercial AM techniques.

TABLE 2.2

Additive Manufacturing vs Traditional Manufacturing

Criteria	Additive manufacturing	Traditional manufacturing
Production cost	For small and medium scale production, production cost is low	Production cost is high for small scale production, however, for mass production these methods are considerably cheaper
Production times	AM manufactures parts directly from CAD data, so there is no need for retooling, supply chain, inventory, etc. which helps in saving time, thus production time is less	Production time is long as these processes depend on availability of tools, molds, inventory, etc.
Production scale	Suitable for low production scale	Suitable for mass production
Design complexity	Most suitable for designing of new products and developing complex shapes	Comparatively less suitable
Material wastage	Minimum or no material wastage	Much material is wasted owing to subtractive operations
Repair work	More suitable for rapid repair/maintenance work	Comparatively less suitable
Raw materials	Many materials, such as some metal alloys, remain unexplored	Suitable for almost all types of materials
Standardization	Standardization of AM needs substantial consideration	Well established technologies
Post-fabrication processing	Depends on type of AM technique being used and varies from little to considerable post-fabrication processing	Needs some kind of post-processing
Resources needed	Only adequate quantity of resources are needed	Large quantity of resources are needed
Prototyping	Most suitable for developing prototypes	Generally not suitable for developing prototypes
On-demand manufacturing	Highly suitable for on-demand or customized manufacturing	Comparatively less suitable for on-demand manufacturing

Source: [12–14]. Joshi and Sheikh, 3D printing in aerospace and its long-term sustainability. *Virtual and Physical Prototyping,* 2015, 10(4): pp. 175–185; 13. Attaran, The rise of 3-D printing: the advantages of additive manufacturing over traditional manufacturing. *Business Horizons,* 2017, 60(5): pp. 677–688; and Groover, *Fundamentals of modern manufacturing: materials, processes, and systems,* 2002, John Wiley & Sons.

References

1. Tofail, S. A. M., Koumoulos, E. P., Bandyopadhyay, A., Bose, S., O'Donoghue, L., Charitidis, C., Additive manufacturing: scientific and technological challenges, market uptake and opportunities. *Materials Today*, 2018, **21**(1): pp. 22–37.
2. Petrovic, V., Haro Gonzalez, J. V., Ferrando, O. J., Gordillo, J. D., Blasco Puchades, J. R., Portolés Griñan, L., Additive layered manufacturing: sectors of industrial application shown through case studies. *International Journal of Production Research*, 2011, **49**(4): pp. 1061–1079.
3. *Additive manufacturing study shows cuts in material consumption and reduced CO2 emissions*, 2013. Available from: www.metal-am.com/additive-manufacturing-study-shows-cuts-in-material-consumption-and-reduced-co%E2%82%82-emissions, accessed January 22, 2019.
4. Ducham, S., *GE asks maker community to push the boundaries of what is 3D printable, launching open engineering quests*, 2013. Available from: www.marketwatch.com/press-release/ge-asks-maker-community-to-push-the-boundaries-of-what-is-3d-printable-launching-open-engineering-quests-2013-06-11, accessed June 11, 2013.
5. Goh, G. D., Agarwala, S., Goh, G. L., Dikshit, V., Sing, S. L., Yeong, W. Y., Additive manufacturing in unmanned aerial vehicles (UAVs): challenges and potential. *Aerospace Science and Technology*, 2017, **63**: pp. 140–151.
6. Pan, Y., White, J., Schmidt, D. C., Elhabashy, A., Sturm, L., Camelio, J., Williams, C., Taxonomies for reasoning about cyber-physical attacks in IoT-based manufacturing systems. *International Journal of Interactive Multimedia and Artificial Intelligence*, 2017, **4**(3).
7. Finnie, I. A. T., Dornfeld, D. A., Eagar, T. W., German, R. M., Jones, M. G., *Unit manufacturing process: issues and opportunities in research*, 1995, National Academy of Sciences, Washington, DC.
8. Huang, S. H., Liu, P., Mokasdar, A., Hou, L., Additive manufacturing and its societal impact: a literature review. *The International Journal of Advanced Manufacturing Technology*, 2013, **67**(5): pp. 1191–1203.
9. Busachi, A., Erkoyuncu, J., Colegrove, P., Martina, F., Watts, C., Drake, R., A review of additive manufacturing technology and cost estimation techniques for the defence sector. *CIRP Journal of Manufacturing Science and Technology*, 2017, **19**: pp. 117–128.
10. Hopkinson, N., Dicknes, P., Analysis of rapid manufacturing—using layer manufacturing processes for production. *Proceedings of the Institution of Mechanical Engineers, Part C: Journal of Mechanical Engineering Science*, 2003, **217**(1): pp. 31–39.
11. Ruffo, M., Hague, R., Cost estimation for rapid manufacturing simultaneous production of mixed components using laser sintering. *Proceedings of the Institution of Mechanical Engineers, Part B: Journal of Engineering Manufacture*, 2007, **221**(11): pp. 1585–1591.
12. Joshi, S. C., Sheikh, A. A., 3D printing in aerospace and its long-term sustainability. *Virtual and Physical Prototyping*, 2015, **10**(4): pp. 175–185.
13. Attaran, M., The rise of 3-D printing: the advantages of additive manufacturing over traditional manufacturing. *Business Horizons*, 2017, **60**(5): pp. 677–688.
14. Groover, M. P., *Fundamentals of modern manufacturing: materials, processes, and systems*, 2002, John Wiley & Sons.

3

Additive Manufacturing Processes

3.1 Introduction

Today's business markets are globally competitive and customer-oriented and are faced with alarming rates of obsolescence. This in turn calls for a need to customize parts and save costs by the manufacturing sector to achieve constant business improvement to stay progressive. AM is around three decades old and has been consistently undergoing much technological up-gradation. It possesses proven versatility in visualizing design and enabling manufacturing advancements. It has attained tremendous significance over recent times. Various terms (for instance, additive, layered, rapid prototyping/tooling/manufacturing; digital fabrication or mock-up; direct or indirect prototyping, tooling and manufacturing; three-dimensional printing or modelling; desktop, solid freeform, generative, on-demand, manufacturing/technology) are frequently used for AM processes. AM has currently established itself as a critical global manufacturing pillar along with conventional manufacturing. In the AM process, material is joined in layers to obtain components directly from CAD data. According to ASTM standards, AM involves joining raw material in layers for fabricating parts as compared to the conventional strategy of material subtraction from bulk. Almost all AM systems utilize similar principles, with a few exceptions of voxelization or hybrid techniques. A variety of ways of classifying AM processes have been suggested by different researchers and standardization associations. A detailed discussion on the same is presented in this chapter.

This is the third chapter of Section A of this book, which presents general details of AM processes. In the previous chapter, a comparison of AM technology with various conventional manufacturing methods has been given in detail. This chapter provides a detailed overview of the classification of AM processes on various bases, such as physical state of raw material, processing techniques, underlying technology, fabrication technique, energy source, raw materials being used, material delivery system. It further discusses common AM defects. Many commercially available AM processes are discussed along various parameters. This chapter concludes with a summary.

3.2 Classification of AM Processes

AM techniques can be classified according to diverse bases. One of these is the basis of the physical form of bulk materials, i.e. liquid, powder and solid sheets, as shown in Figure 3.1. AM processes are most commonly classified on this basis. Second, the mechanism employed for transferring STL data into physical structures is another parameter based upon which AM processes can be of four types: 1D (one-dimensional), multiple 1D, array of 1D and 2D channels. This is shown in Table 3.1. Combining both the aforementioned bases together forms a better way of streamlining AM classification and is presented in Table 3.2. An interesting observation on this array are the vacant spaces, which provide a hint to AM personnel regarding the domain of probable future research/advances. The next classification can be with respect to working method or essential technology, as shown by Figure 3.2.

Another classification finds its basis in the energy source used to join materials, for example, laser, electron beams, etc. The next method of AM classification is on raw materials being used, such as ceramics, powders, plastics and so on. The seventh basis is material delivery system as depicted in Table 3.3. Eighth way of classifying is based upon ASTM F42 Committee guidelines which is presented in Table 3.4. A detailed description of the ASTM-based classification is presented in Table 3.5.

An example of comparison of these seven AM categories on the basis of use of relative energy with respect to the speed and resolution is presented in Figure 3.3.

The bases of classification of AM processes discussed above are listed below for easy reference of readers:

1. Physical raw material state
2. Transfer of .stl data into deposition method
3. Combination of 1 and 2
4. Working methodology/ underlying technology
5. Energy source
6. Raw materials
7. Material delivery system
8. ASTM guidelines.

TABLE 3.1

AM Classification Based upon Mechanism Employed for Transferring .stl Data into Physical Structures

One-dimensional channel	Two one-dimensional channels	Array of one-dimensional channels	Two-dimensional channels
SLA, SLS, etc.	Dual beam SLA, LST, etc.	Objet, 3D printing, etc.	Envisiontech Micro TEC, DPS, etc.

Adapted from: [1, 2]. Pham D. T., Dimov, S. S., *Rapid manufacturing: the technologies and applications of rapid prototyping and rapid tooling*, 2001, Springer; and Gibson, I., Rosen, D. W., Stucker, B., Additive manufacturing technologies, in *Rapid Prototyping to Direct Digital Manufacturing*, 2010, Springer U.S, pp. 1–459.

TABLE 3.2

AM Classification Based upon Bulk Materials along with Manner of Transfer

State of raw material	Mechanism of transfer			
	One-dimensional	Multi-one-dimensional	One-dimensional array	Two-dimensional
Liquid	Stereo lithography, liquid thermal polymerization		Objet quadra process	Solid ground curing, rapid micro product development
Discrete particles	Selective laser sintering, laser sintering technology, laser engineering net shaping, laser assisted chemical vapor deposition, selective laser reactive sintering, gas phase deposition, selective area laser deposition	Laser sintering technology	Three-dimensional printing	Direct photo shaping
Molten material	Fused deposition modelling, ballistic particle manufacture, three-dimensional welding, precision droplet based net-form manufacturing		Multi-jet modelling	Shape deposition modelling
Solid sheets	Laminated object manufacturing, paper lamination technology			Solid foil polymerization
Electroset fluids				Electro-setting

Source: [3]. Srivastava, *Some studies on layout of generative manufacturing processes for functional components*, 2015, Delhi University (own thesis).

TABLE 3.3

Classification on Basis of Material Delivery System

Description	Powder bed	Powder feed
Deposition strategy	Uses a powder deposition system with coating mechanism to spread layers of powder onto a substrate plate and powder reservoir	Material flows through a nozzle being melted from a beam right on the surface of treated part
Layer thickness	Produces 20–100-micron size layers that can either be bonded together using lasers or stuck together using melting	Highly precise systems based on layer deposition thickness between 0.1 mm to several centimetres
Features	Choice of process depends upon: (a) laser unit; (b) powder handling; (c) built chamber size, etc.	Cladded substance has metallurgical bonding with base material and undercutting is absent. Process is not same as other welding techniques
Examples	(a) Selective laser sintering; (b) laser cusing; (c) direct metal laser sintering; (d) electron beam welding	(a) Laser cladding; (b) laser metal deposition; (c) directed energy deposition; (d) laser engineered net shaping
Commercial systems	(a) Arcam AB; (b) Matsuura; (c) Hogonas	(a) BeAM; (b) Trumpf; (c) Sciaky

Source: [4]. Rathee et al., *Friction based additive manufacturing technologies: principles for building in solid state, benefits, limitations, and applications*, 1st ed., 2018, CRC Press, Taylor & Francis Group.

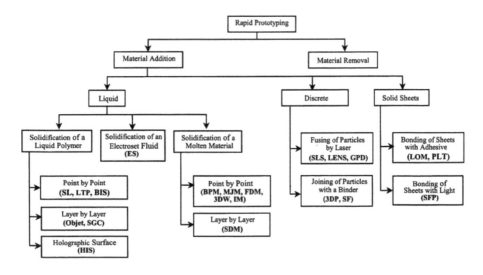

FIGURE 3.1

AM Classification Based upon Raw Material State. (*Source:* [1]. Reproduced from Pham and Dimov, *Rapid manufacturing: the technologies and applications of rapid prototyping and rapid tooling*, 2001, Springer , with permission from Springer Nature.)

TABLE 3.4

Classification on Basis of ASTM F42 Committee Guidelines (Materials and Technology Utilized)

Description	Metallic machines	Non-metallic machines
Technology	Powder bed fusion (PBF); selective laser melting (SLM); electron beam melting (EBM); direct energy deposition (DED) Laser vs. e-beam	Material extrusion, material jetting, vat photopolymerization
Feedstock	Metallic feedstock	Non-metallic feedstock, e.g. powder, resins, plastics, glass
Physical state	Solid (powder, sheet or wired form)	Solid (powder, sheet, wire), liquid or gas
State of consolidation	Consolidation of feedstock into parts of full density	Cannot produce full density components

Source: [4]. Rathee et al., *Friction based additive manufacturing technologies: principles for building in solid state, benefits, limitations, and applications*, 1st ed., 2018, CRC Press, Taylor & Francis Group.

TABLE 3.5

Description of Individual Techniques on Basis of ASTM F42 Committee Guidelines

Serial no.	Technique	Description
1	VAT photopolymerization	Utilizes liquid photopolymer resin vat from which model is fabricated in a layered fashion
2	Material jetting	Creates objects in a fashion similar to a two-dimensional inkjet printer where material is jetted onto a build platform utilizing a continuous/DOD approach
3	Binder jetting	Utilizes two materials, one powder-based material and the other binder (liquid). Fabrication is accomplished by a printhead depositing layers
4	Material extrusion	FDM is an example of material extrusion process where material is drawn through a nozzle in heated form and deposited in layers. The nozzle can move horizontally and a platform moves up and down vertically after each new layer is deposited
5	Powder bed fusion	Includes following commonly used printing techniques: direct metal laser sintering (DMLS), electron beam melting (EBM), selective heat sintering (SHS), selective laser melting (SLM) and selective laser sintering (SLS)
6	Sheet lamination	Includes ultrasonic additive manufacturing (UAM) and laminated object manufacturing (LOM). UAM process uses sheets or ribbons of metal, which are bound together using ultrasonic welding
7	Directed energy	Directed energy deposition (DED) covers a range of terminology: laser engineered net shaping, directed light fabrication, direct metal deposition, 3D laser cladding. It is a more complex printing process commonly used to repair or add additional material to existing components

Source: [4]. Rathee et al., *Friction based additive manufacturing technologies: principles for building in solid state, benefits, limitations, and applications*, 1st ed., 2018, CRC Press, Taylor & Francis Group.

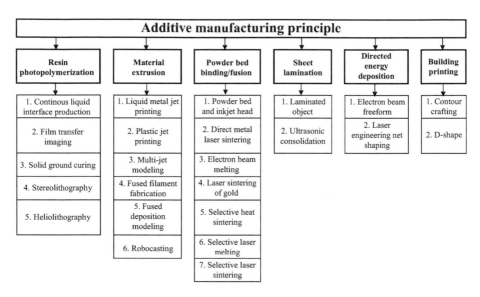

FIGURE 3.2
Classification on the Basis of Working Principle. (*Source:* [3, 4]. Reproduced from Pham and Dimov, *Rapid manufacturing: the technologies and applications of rapid prototyping and rapid tooling,* 2001, Springer , with permission from Springer Nature; Rathee et al., *Friction based additive manufacturing technologies: principles for building in solid state, benefits, limitations, and applications,* 1st ed., 2018, CRC Press, Taylor & Francis Group.)

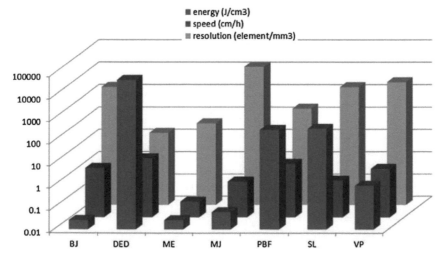

FIGURE 3.3
Comparison of Seven AM Categories on the Basis of Relative Energy Use with Respect to Speed and Resolution. (*Source:* Reprinted from [5], Tofail et al., Additive manufacturing: scientific and technological challenges, market uptake and opportunities. *Materials Today,* 2018, 21(1): pp. 22–37, under Creative Commons Licence. Adapted from [6], Lee, et al., Fundamentals and applications of 3D printing for novel materials. *Applied Materials Today,* 2017, 7: pp. 120–133.)

3.3 Common AM Processes

A non-exhaustive summary of a few AM processes is presented in Table 3.6 as an attempt to introduce prominent commercial AM techniques to readers.

3.4 Summary

AM techniques are being increasingly utilized in aviation, automotive and marine sectors due to their ability to fabricate intricate parts. They are increasingly replacing conventional manufacturing techniques for numerous critical applications. However, despite being the forbearer of technological advancements, these technologies are faced with multifold challenges.

In order to successfully utilize them, overcoming these challenges is necessary to open novel avenues for AM applications. Presently, even the MAM techniques inherently face challenges due to occurrence of shrinkage of cavities, porosity, inclusion of foreign particles in molten metal, solidification defects because of liquid-solid phase transformation and so on. When the microstructures of MAM parts are examined, they exhibit non-homogeneous microstructures and anisotropic mechanical characteristics which restrict the utility of AM fabricated parts for critical applications. Also, the production quantum is restrained due to limited build volumes and need for close control of build chambers in the majority of fusion-based AM methods These challenges can be easily overcome by using friction-based AM techniques. However, even these friction-based techniques have their own set of drawbacks and limitations.

In general, it can be concluded that, though AM techniques are game-changers and enabling technologies, much research work is required to fully utilize their potential for critical applications apart from known commercial applications. A detailed cost benefit and economic analysis is therefore mandatory before choosing a particular process. Also, the intricacies of various AM processes need to be clearly understood before deciding upon the suitability of a particular process for any application. In this direction, this chapter discusses the various bases of classification and prominent AM processes.

In this chapter, AM techniques are classified on various bases. Various commercial AM techniques are introduced and their details are presented. The readers are also introduced to various AM defects. The next chapter presents details of the historical evolution of AM techniques. It also aims to provide a tentative historical timeline of the AM processes.

TABLE 3.6

Summary of Commercial AM Processes

S. no.	Technology	Raw materials	Key issues	Strengths	Applications
1	Selective laser sintering (SLS)	Either coated powders or powder mixtures, polymers, polystyrene, metals (steel, titanium, etc.), alloy mixtures, green sands	(a) Involves use of expensive and potentially dangerous lasers (b) Components have porous surfaces	(a) Fully self-supporting so nesting is very easy (b) Parts offer high strength, stiffness and chemical resistance	Physical models of aircrafts, used specially where small number of high-quality parts are required
2	Solid ground curing	Photosensitive resins	Produces large amount of waste, high operating cost, complex system	Does not require support structure (as wax fills the void), good accuracy, high fabrication rates	Models, prototypes, patterns, production parts
3	Inkjet printing (IJP)	Wax-like thermoplastic materials		(a) Clean and low-priced concept modellers (b) Suited for small, accurate and intricate parts	Suitable for concept modelling and investment casting models
4	Fused deposition modelling (FDM)	Thermoplastics, nylon, ceramics, metals, etc.	(a) Slow speed leading to large build time (b) Anisotropy (c) Less surface finish	Cost, material	Form and functional models, concept models, prototypes, medical models
5	Stereolithography (SL)	Liquid resins which can be hardened using photopolymerizing effect	(a) Process controlled by laser quality, scanning periods, resin recoating (b) Support structures required (c) Portability issues due to liquids, post-processing problems	One of oldest AM techniques; underwent much research and emphasis during initial phases; Sufficiently large component can be made accurately	Now obsolete

6	Laminated object manufacturing (LOM)	Paper, wood, plastic sheets	(a) Dimensional accuracy less than SLA and SLS (b) Limited scope of material, surface finish and accuracy issues	(a) No milling necessary (b) Low cost due to readily available raw materials (c) Large components can be handled	Rapid tooling, prototyping, pattern making, medical applications
7	Laser engineered net shaping (LENS)	Metallic powder of steels, nickel-based super alloys, Inconel, titanium, cobalt	(a) Severe overhangs (b) Solidification microstructures (c) Mechanical properties (d) Surface finish (e) Secondary finishing process required (f) Support structure problem	(a) Can use composite powder mixture (b) Lowers cost and time requirement (c) Parts fully dense with non-degraded microstructures	Repair, overhaul, production for aerospace, defense and medical markets
8	Laser additive manufacturing (LAM)	Steels, nickel-based super alloys, Inconel, titanium, cobalt	(a) Severe overhangs (b) Solidification microstructures (c) Mechanical properties	(a) Can use composite powder mixture (b) High cooling rates	Aerospace, defense, automotive and biomedical industries
9	Laser sintering	Multi-component powders composed of high melting point component called structural metal, low melting point component called binder and flux/deoxidizing agents called additives	(a) Inert gas environment required (b) High-quality laser systems are required, which also greatly affect powder consolidation (c) Mushy zone due to solid liquid wetting	(a) Can produce multi-material alloys (b) Appreciably good mechanical properties can be obtained by controlling SLPS	Medical models, metallic molds, etc.
10	Laser melting	Multi-material ferrous and non-ferrous powders	(a) Requires high-quality laser systems (b) Requires high energy levels (c) Instability of melting points need attention	(a) Produces fully dense components (b) Mechanical characteristics comparable to bulk materials (c) Can process non-ferrous materials	Pure non-ferrous metals full density components with high strength like titanium, multi-material components, etc.

Continued

TABLE 3.6 *Continued*

S. no.	Technology	Raw materials	Key issues	Strengths	Applications
11	Laser metal deposition	Multi-material powders	(a) Requires high-quality specially designed coaxial feeder system (b) Requires high-quality laser systems (c) Costly and intricate (d) Requires intricate closed loop control	(a) Quite advanced (b) Multi-material delivering capabilities (c) Patented closed loop controls	Manufacturing new components for repair and rebuilding of worn out/damaged products, for wear and corrosion-resistant coatings, etc.
12	Digital metallization	Metal foil, plastics, paper	(a) Expensive (b) Few modellers available (c) Suitable for late-stage customization	(a) Can be used to produce design of different colors and light effects (b) Suitable as cost-effective customization technology to implement metallic effects	Decorative items, functional items like automobiles, electromagnetic shielding, circuit paths, etc.
13	Prometal	Iron, bronze, glass, etc.	(a) Low speeds (b) Limited volumes (c) Technological and economical limitations	(a) Offers customized hardware in much less time (b) Suitable for both metals and glasses	3D printed door pulls, knobs, knockers, decorative items, customized 3D hardware components
14	Direct metal deposition (DMD)	Steel, Waspalloy, titanium	(a) Post-processing (b) Low build rate (c) Complex process	(a) Neutral gas (b) Precision part pick up (c) Quick tool path operation	Used to repair and rebuild worn/damaged parts, tools, dies, cutters, etc.
15	Selective laser melting (SLM)	Steel, Inconel, titanium, cobalt, aluminium	(a) Post-processing (b) High production cost (c) Properties	Complex geometry is achievable	Medical and dental applications, lightweight structures, heat exchangers

#	Process	Materials	Disadvantages	Advantages	Applications
16	Easy clad	Steel, Waspalloy, titanium	(a) Post-processing (b) Low build rates	(a) Neutral gas (b) Better properties in comparison to castings	Construction of parts for aerospace and aircraft industry
17	Ultrasonic additive manufacturing (UAM)	Plastic or metallic sheet like Aluminium alloys	(a) Difference in microstructure at interfaces and non-interfaces (b) Weak link in build (c) Foil preparation (d) Low build volume (e) Mechanical properties	(a) Solid state processing method (b) Multi-material structures (c) Environment friendly (d) Involves less heat so no melting hence no distortion (e) Good bonding properties	Injection moulding dies, parts with embedded channels, incorporation of second phase materials, intricate geometry components, etc.
18	Direct metal laser sintering	Steels, Al, cobalt, titanium	(a) Post-processing (b) High operating cost	Multi-material structures	Turbine blades
19	Laser Cusing	Precious metals, steels, titanium, aluminium	(a) Mechanical properties (b) Low build rates	(a) High-quality finishing (b) Reduction in stresses	Dental, jewellery automotive, aerospace, medical devices, tool and mold manufacture
20	Electron beam melting (EBM)]	Fine metallic powders of copper, beryllium, steels, titanium, aluminium, nickel, etc and alloys	(a) Surface quality (b) Solidification defects (c) High skill requirement (d) Large power requirement	(a) Faster build compared to DMLS and SLM (b) Delivers mechanical strength with less mass, cost and weight is reduced	Small series parts, prototypes, support parts (jigs, fixtures, etc.)
21	Shaped metal deposition	Metallic wires	(a) Poor dimensional accuracy and surface finish (b) Environmental hazards (c) High cost	(a) High production rates (b) Lower costs (c) Shorter lead times (d) High deposition rates and efficiencies (e) Denser part production capability	Features with intricacy and complexity, large-scaled parts (ex-aerospace and metal-die)

Continued

TABLE 3.6 *Continued*

S. no.	Technology	Raw materials	Key issues	Strengths	Applications
22	Electro discharge deposition	Any material which is conductive can be utilized as raw material	Magnetic field needs to be very carefully controlled as it has major impact over deposition	(a) Used to manufacture micro products (b) Integration of multitude of functions is possible	Miniatured products with high-aspect ratio micro feature
23	Selective layer chemical vapor deposition	Metals like aluminium, copper, tungsten, etc.	(a) Controlling chemistry of vapor (b) Choosing suitable vapor deposition technique	(a) Conformal coverage of irregularly shaped surfaces (b) High throughput	Thin film applications, functionally gradient materials
24	Shape deposition modelling	Powder (stainless steel invar, aluminium, titanium)	(a) High cost (b) Support material requires manual work for removal	(a) Wide variety of material (b) Use of conventional machine tools (c) No stair-step effect unlike others	Dense components with accuracy, excellent surface finish and metallurgical bonding

Source: [2, 4, 7]. Gibson et al., Additive manufacturing technologies, in *Rapid Prototyping to Direct Digital Manufacturing*, 2010, Springer U.S., pp. 1–459; Rathee et al., *Friction based additive manufacturing technologies: principles for building in solid state, benefits, limitations, and applications*, 1st ed., 2018, CRC Press, Taylor & Francis Group; and Srivastava et al. A review on recent progress in solid state friction based metal additive manufacturing: friction stir additive techniques. *Critical Reviews in Solid State and Materials Sciences*, 2018: pp. 1–33.

References

1. Pham D. T., Dimov, S. S., *Rapid manufacturing: the technologies and applications of rapid prototyping and rapid tooling*, 2001, Springer.
2. Gibson, I., Rosen, D. W., Stucker, B., Additive manufacturing technologies, in *Rapid Prototyping to Direct Digital Manufacturing*, 2010, Springer U.S, pp. 1–459.
3. Srivastava, M., *Some studies on layout of generative manufacturing processes for functional components*, 2015, Delhi University.
4. Rathee, S., Srivastava, M., Maheshwari, S., Kundra, T. K., Siddiquee, A. N, *Friction based additive manufacturing technologies: principles for building in solid state, benefits, limitations, and applications*, 1st ed., 2018, CRC Press, Taylor & Francis Group.
5. Tofail, S. A. M., Koumoulos, E. P., Bandyopadhyay, A., Bose, S., O'Donoghue, L., Charitidis, C., Additive manufacturing: scientific and technological challenges, market uptake and opportunities. *Materials Today*, 2018, **21**(1): pp. 22–37.
6. Lee, J.-Y., An, J., Chua, C. K., Fundamentals and applications of 3D printing for novel materials. *Applied Materials Today*, 2017, **7**: pp. 120–133.
7. Srivastava, M., Rathee, S., Maheshwari, S., Siddiquee, A. N., Kundra, T. K., a review on recent progress in solid state friction based metal additive manufacturing: friction stir additive techniques. *Critical Reviews in Solid State and Materials Sciences*, 2018: pp. 1–33.

4

Evolution of Additive Manufacturing Technologies

4.1 Introduction

Today's concept of additive manufacturing technologies emerged from the term "rapid prototyping" (RP) that was widely used for reduction of process time to facilitate rapid creation of parts which were to be subsequently fabricated and commercialized. It involves development of an idea/software/part, that may either have manufacturing or management functionality, in a piecewise manner, directly from digital inputs, which allows several design iterations and testing before full-fledged manufacturing. However, over time this definition is becoming rather partial for users since these technologies give outputs closer to the final products and labelling all of them merely as prototypes would be inaccurate. The term RP is also non-exhaustive in the sense that it does not include the "additive" capability and tendency of these processes. It would therefore be inaccurate and an undervaluation of these techniques if we still continue to use the term "RP." The term "additive manufacturing" (AM) would therefore be a better substitute for RP. A standard definition of AM techniques was also required and the ASTM committee bridged this gap. ASTM gave the following definition of AM: "A process of joining materials to make objects from 3D model data, usually layer upon layer, as opposed to subtractive manufacturing methodologies." Synonyms of AM are: "additive fabrication, additive processes, additive techniques, additive layer manufacturing, layer manufacturing, and freeform fabrication" [1]. It is noteworthy that the use of term "AM" has been advocated by ASTM committee consensus. Thus, AM can be understood as a process possessing the capability of producing intricate final products with negligible process planning directly from three-dimensional CAD data. AM processes require only detailed dimensions, materials and machine working details, in contrast to the conventional processes requiring detailed engineering drawings, their analysis, tools, jigs, fixtures, etc.

Today, there is a wide variety of machines available under the umbrella of AM. However, depending upon varied AM strategies, outputs obtained vary significantly and can even conflict. Therefore, standardization is extremely significant. In 2009, ASTM International formed the F42 committee that aims to provide standardization of material, process, terms, designs, exchange format as well as testing techniques. Standards such as ASTM F2792-12a, ISO/ASTM DIS 52910.2, ISO/TC 261, ISO 17296–4:2014, ISO 10993–1, ASTM F 2129, ASTM 756, ISO 10993–6 help in aspects of selecting, characterizing and testing various AM techniques for different applications. Owing to the tremendous pace of AM growth over the last decade, there is need for newer standardization methods. Also, better terminology is being introduced by competent authorities on a regular basis.

To fully understand AM technology, it is important to understand the historical development and timeline of AM processes as discussed in detail over the subsequent sections.

This is the fourth chapter of Section A of this book, which presents general details of AM processes. In the previous chapter, a detailed classification of AM techniques on various bases has been provided. A discussion of various AM defects is also presented, in addition to a summary of various commercially available AM techniques. This chapter aims to provide a detailed account of the evolution of AM technology over time. It also presents a detailed timeline of the AM processes and concludes with a summary.

4.2 Evolution of AM Technologies

Development of AM technologies was gradual over the first decade of its advent. However, recently owing to advancements in manufacturing, associated technologies, customer expectations and global markets, the development rates have witnessed a huge upsurge. The first AM patent was filed by Dr. Kodama (Japan, 1980). Dr. Charles Hull (SLA, 1987) is considered to be the forbear of the effective commencement of AM/3DP. Almost around the same time, a patent by Carl Deckard (SLS, 1987) was filed and granted, and later acquired by 3D Systems. Dr. Scott Crump (FDM, 1989) filed the next patent which was granted, and later acquired by Stratasys (1992). This was a landmark development in the sense that a vast variety of RepRap modellers are based upon FDM technology. EOS GmbH (Hans Langer, 1989) was founded which focussed largely on laser sintering, stereos and direct metal laser sintering systems. Several other systems emerged around the same time; these included laminated object manufacturing (LOM) by Helisys (Michael Feygin, 1990s), ballistic particle manufacture (BPM) (William Masters, 1990s), solid ground curing (SGC) by Cubital (Ifz Chak

Pomerant et al., 1990s), 3DP by 3D systems (Emaneul Sachs,1990s). Sanders Prototype (1996), Z Corp (1996) and Objet (1998) also came into being. LOM, SLA, BPM and SGC have not survived over the years due to various technical limitations. One remarkable event during the middle of the twentieth century needs special mention. This included development of two discrete categories of AM processes: first, high-end expensive AM systems for fabricating complex, intricate and high-value parts to be used in aerospace, automobile, medicine, jewellery, etc. applications; second, a moderately priced category of concept modellers to be used in concept modelling, functional prototyping, assembly validation, etc. applications. The second category of modellers emerged due to the cut-throat highly competitive markets.

The twenty-first century saw a shift in the AM applications paradigm from R&D and prototyping to industrial and functional part fabrication. The term RP has gradually been replaced by AM, which includes both direct and indirect AM and involves a variety of aspects of prototyping, tooling and manufacturing. Emergence of technologies like selective laser melting (SLM) (2000), Envision Tec (2002), EBW/Exone (2005) was witnessed over time. Gradually, due to a variety of favorable features like improved prices, speeds, materials and accuracy, 3D Systems developed their first AM system priced below $10,000. Next in line came the concept of a $5000 desktop factory system that had to be shelved and did not come into being. Self-replicating machines based upon RepRap technology (Dr. Bowyer, 2009) led to a revolution in the AM market and led to this open source technology transforming itself into an industrial practice through which every interested person could access the idea of AM printers. Since then a large variety of AM printers based upon this technology have become widely available. Alternative AM processes emerged via B9 creators (2012). Form 1 Modeler (2012) also materialized around the same time. Over the last decade, progress in the field of AM has been so widespread that it is difficult to single out every noteworthy advancement. However, in an attempt to demonstrate major developments in field of AM, a detailed timeline of AM is presented in the next section.

4.3 Timeline of AM Technology

AM technology has grown leaps and bounds over last three decades and today it possesses the ability to complement as well as substitute conventional manufacturing in a wide variety of ways. Today numerous different AM techniques are known. With an aim to introduce readers to the various AM techniques, a non-exhaustive timeline of these AM technologies and modellers is presented in Table 4.1.

TABLE 4.1

Timeline of AM Processes

Sr. no.	Year of advent	Name of process	Patented/ commercialized by
1	1987	Stereolithography	3D systems
2	1991	Fused deposition modelling	Stratasys
		Solid ground curing	Cubital
		Laminated object manufacturing	Helisys
3	1992	Selective laser sintering	DTM (now part of 3D Systems)
		Soliform stereolithography system	DuPont/Teijin Seiki
4	1993	Direct shell production casting	Soligen
		Denken's SL system	Denken
		QuickCast	3D systems
5	1994	ModelMaker	Solidscape/Sanders Prototype
		Meiko's stereolithography machine	Meiko
		Solid center-laminated object manufacturing	Kira Corp
		Fockele & Schwarze (F&S) stereolithography machine	Fockele & Schwarze (F&S)
		EOSINT	EOS
6	1995	Ushio stereolithography machine	Ushio
7	1996	Genisys machine	Stratasys
		3D printer – Actua 2100	3D Systems
		3D printer – Z402	Z Corp.
		Semi-automated paper lamination system	Schroff Development
		Personal Modeler 2100 – ballistic particle manufacturing	BPM Technology
		DuPont's Somos stereolithography technology	Aaroflex
		Zippy paper lamination systems	Kinergy
8	1997	Laser additive manufacturing	AeroMet
9	1998	Beijing Yinhua Laser Rapid Prototypes Making and Mould Technology	Tsinghua University
		E-DARTS stereolithography system	Autostrade
		Laser-engineered net shaping	Optomec
10	1999	ThermoJet	3D Systems
		ProMetal RTS-300 Machine	Extrude Hone AM business
		Steel powder-based selective laser melting system	Fockele & Schwarze
		Controlled metal build-up (CMB) machine	Röders
11	2000	Rapid Tool Maker	Sanders Design International
		Color 3D printer	Buss Modeling Technology
		Quadra	Objet Geometries
		PatternMaster	Sanders Prototype – Solidscape
		Direct metal deposition	Precision Optical Manufacturing
12	2001	Ultrasonic consolidation process	Solidica
		Perfactory machine based on digital light processing	Envisiontec
		EuroMold 2001	Concept Laser GmbH

Sr. no.	Year of advent	Name of process	Patented/ commercialized by
13	2002	Direct metal deposition machine	POM
14	2003	Z Printer310	Z Corp
		Sony stereolithography machines	Sony Precision Technology
		T612 system	Solidscape
		InVision 3D printer	3D Systems
		Low-cost Wizaray stereolithography system	Chubunippon
		EOSINT M 270	EOS
		TrumaForm LF and DMD505	Trumpf
15	2004	Three versions of FDM Vantage	Stratasys
		Vanquish photopolymer-based system	Envisiontec
		RX-1 metal-based machine	ProMetal division of Ex One
		InVision HR, Sinterstation HiQ	3D Systems
		Ultrasonic consolidation system – Formation machine	Solidica
		Dual-vat Viper HA stereolithography system	3D Systems
		Vero FullCure 800 series	Objet
		EOSINT P 385	EOS
		M1 cusing laser-melting machine	Concept Laser
		DigitalWax 010 and 020 systems	Next Factory (now DWS)
		T66 and T612 Benchtop systems	Solidscape
16	2005	Spectrum Z510	Z Corp.
		Sinterstation Pro	3D Systems
		InVision LD	Solido and rebranded by 3D Systems
		SEMplice laser-sintering machine	Aspect Inc.
		Eden500V	Objet Geometries
		Zprinter 310 Plus	Eden500V
		Large Viper Pro SLA	3D Systems
		Sand-based Sprint machine	Ex One's ProMetal division
		SLM ReaLizer 100	MCP Tooling Technologies
		VX800 machine	Voxeljet Technology GmbH
17	2006	Eden350/350V platform. Eden250 3D	Objet Geometries
		InVision DP	3D Systems
		Vantage X systems	Stratasys
		Dimension 1200 BST and SST	Stratasys
		NanoTool	DSM Somos
		Zscanner 700 handheld 3D scanner	Z Corp.
		Formiga P 100 laser-sintering system, EOSINT P 390 and EOSINT P 730	EOS
		VX800 machine	Voxeljet Technology
		Small Perfactory Desktop System	Envisiontec
		SLM ReaLizer 100 selective laser-melting machine	MTT
18	2007	V-Flash 3D printer	3D Systems
		Dimension Elite 3D printer	Stratasys
		D66 and R66 T66 machines	Solidscape
		ZPrinter 450	Z Corp.
		A2 electron beam melting (EBM) machine	Arcam
		Formiga P 100 laser-sintering system	EOS

Continued

TABLE 4.1 *Continued*

Sr. no.	Year of advent	Name of process	Patented/ commercialized by
		FDM 200mc machine	Stratasys
		M2 cusing system	Concept Laser
		FDM 400mc	Stratasys
		VX500 system	Voxeljet
		FDM 900mc	Stratasys
		PerfactoryXede	Envisiontec
		Connex500 3D-printing system	Objet Geometries
		FigurePrint approach	Microsoft
		Pollux 32 selective mask-sintering system	Sintermask
19	2008	FDM 360mc	Stratasys
		ProJet HD3000	3D Systems
		Dimension 1200es 3D printer	Stratasys
		V-Flash desktop modeller	3D Systems
		T76 precision waxprinting System	Solidscape
		LENS MR-7 machine	Optomec
		iPro 9000 SLA Center stereolithography system	3D Systems
		ProJet SD 3000 3D printer	3D Systems
		PolyStrata microfabrication technology	Nuvotronics
		TetraLattice Technology	Milwaukee School of Engineering
		iPro 9000 XL SLA Center, iPro 8000 MP SLA Center, ProJet CP 300 RealWax 3D printer, large-format ProJet 5000, sPro 140 and 230 SLS Centers	3D Systems
		SLM 250-300	MTT
		Matrix 3D printer	Mcor Technologies
		MLS MicroLightSwitch technology	Huntsman Advanced Materials
		ZPrinter 650	Z Corp
		Alaris30 PolyJet machine, Eden260V machine	Objet
		Araldite Digitalis	Huntsman Advanced Materials
		EOSINT P 800	EOS
		Fortus Finishing Stations	Stratasys
		Spore Sculptor facility	Electronic Arts, Z Corp.
		JuJups.com	Genometri of Singapore
		Eden260V machine	Objet
20	2009	ASTM Committee F42 on Additive Manufacturing Technologies established	ASTM International headquarters
		uPrint Personal Printer	Stratasys
		Shapeways Shops	Shapeways
		SLM 50	ReaLizer GmbH
		RapMan 3D printer Kit	Bits from Bytes
		Cupcake CNC product	MakerBot Industries
		ULTRA bench-top DLP-based system	Envisiontec
		PreXacto line	Solidscape
		VFlash 3D printer	3D Systems
		Connex350 system	Objet Geometries
		Dimatix Materials Printer DMP-3000	Fujifilm Dimatix

Sr. no.	Year of advent	Name of process	Patented/ commercialized by
		Standard terminology on Additive Manufacturing Technologies	ASTM International Committee F42
		Monochrome ZPrinter 350 machine	Z Corp.
		EOSINT P 395 and EOSINT P 760	EOS
		ProJet 5000	3D Systems
		SD300 Pro system	Solido
		Three DLP-based visible light photopolymer 3D printers	Carima
		VX800HP	Voxeljet
		Solidscape users group	Steven Adler of A3DM
		Design-your-own toy website, MAQET (maqet.com)	Keith Cottingham
21	2010	uPrint Plus	Stratasys
		Aerosol Jet Display Lab System	Optomec
		Cometruejet	Microjet Technology
		EasyCLAD Systems for its laser metal deposition equipment	Irepa Laser
		Dental manufacturing facility	Renishaw plc (UK)
		Consumer-oriented products made from glass	Shapeways
		Began delivery of HP-branded FDM machines	Stratasys
		ZBuilder Ultra	Z Corp.
		Portable personal UP! 3D printer	Delta Micro Factory Corp.
		Monochrome ZPrinter 150 and color ZPrinter 250	Z Corp.
		Wax drop-on-demand printers	Solidscape
		Tool-making process based on direct metal laser sintering	EOS and GF AgieCharmilles
		Service named InvisHands started	INUS
		EOSINT M 280 system	EOS
		Objet24 3D printer	Objet
		HD 3000plus and CPX 3000plus Systems	3D Systems
		ProJet 6000	ProJet
		Axis 2.1 kit, Glider 3.0	BotMill
22	2011	Aerosol Jet printhead, Objet260 Connex	Optomec
		Fortus 250mc	Stratasys
		Released Additive Manufacturing File (AMF) format	Additive Manufacturing Technologies
		RepRap-based Buildatron 1 3D printer	Buildatron Systems
		ProJet 1500	3D Systems
		RepRap-based Solidoodle 3D printer	Solidoodle
		Released standard terminology for coordinate systems and test methodologies	ASTM International Committee F42 and ISO Technical Committee 261
		New line of ultrasonic additive manufacturing	Fabrisonic
		Unveiled its online digital manufacturing and social network	Kraftwurx
		SLM 280 HL	SLM Solutions
		Freeform Pico	Asiga
		ProJet 1000	3D Systems
		LUMEX Avance-25	Matsuura
23	2012	MakerBot Replicator	MakerBot

Continued

TABLE 4.1 *Continued*

Sr. no.	Year of advent	Name of process	Patented/ commercialized by
		Cube	3D Systems
		MAGIC LF600	EasyClad
		Second-generation machine for $499	Solidoodle
		Mojo 3D printer	Stratasys
		Objet30 Pro	Objet
		UP! Mini	Delta Micro Factory Corp.
		ProJet 7000	3D Systems
		EOS M series machines	EOS and Cookson Precious Metals
		Iris full-color, paper-based, sheet lamination system	MCor Technologies
		3Z line of high-precision wax 3D Printers	Solidscape
		Flex Platform	ExOne
		Replicator 2 and 2X platforms	MakerBot Industries
		Matrix 300+ machine	Mcor
		Form 1	Formlabs
		Solidoodle 3D printer for $499	Solidoodle
		Objet1000 system	Objet
		VXC800 machine	Voxeljet
		Formiga P 110 laser-sintering machine	EOS
		ProJet 3500 HDMax and ProJet 3500 CPXMax	3D Systems
		Ultimaker 3D printer	Ultimaker
24	2013	CubeX multihead 3D printer	3D Systems
		3Dent printer	Envisiontec
		Arcam Q10 machine	Arcam
		Replicator 2	MakerBot and Autodesk
		HeartPrint service	Materialise
		ProJet x60 family of machines	3D Systems
25	2014	First powder bed fusion four-laser concept by EOS	
26	2015	First AM part used in jet engine by GE	
27	2016	GE took over Concept laser and Arcam, Oerlikon acquired Citim	
28	2017	DMG MORI presented new powder bed fusion modeller along with Realizer	
29	2018	AM systems ready for series production, simultaneous part creation by four lasers, massively increased built areas, hybrid AM systems based on solid-state processing developed	

Source: [2]. Rathee et al., *Friction based additive manufacturing technologies: principles for building in solid state, benefits, limitations, and applications*, 1st ed., 2018, CRC Press, Taylor & Francis Group.

Table 4.1 presents details of modellers and technologies up to whereas 2013, for the years after 2013 applications and landmark events have been highlighted. This is chiefly because of the shift in the AM paradigm from a technology to applications perspective over the last few years.

4.4 Summary

Caffrey et al. [3] forecasted several AM trends over forthcoming years that include:

1. Expired primary patents will amount to emergence of low/medium cost AM systems possessing enhanced availability corresponding to increased customer demands. The same will hold true for newer and improved raw materials.
2. Speeds of fabrication will significantly improve owing to material/ design advances. This will in turn result in part fabrication times in minutes.
3. Hybrid AM will form the basis of future machines but a reduction in their versatility will take place.
4. Future AM systems will be able to process multi-material parts.
5. Tissue engineering and biomedical applications of AM systems will increase enormously.

Most of the above predictions are found to hold true in the context of present AM developments. Though AM is not meant to fully substitute conventional manufacturing methods, yet there are domains where this technology has unique benefits. A huge growth in the AM sector and in turn appreciable savings of time and cost are being forecast by researchers over forthcoming years; for example, Attran et al. [4] predicted that AM processes will be four times faster and half as costly as their present values by 2023.

The growth trends present a clear picture of these predictions and support most of them. One most crucial change is observed in the consumption pattern and product expectation of customers. Today the desire to customize and innovate products is at its peak, as can be evidenced from the newer manufacturing reforms in terms of markets and practices. A large number of researchers, institutions and industries have embedded AM into their innovative, research and manufacturing habits. Figure 4.1 shows the various opportunities in terms of technical and operational aspects in the field of AM.

AM is currently used for numerous applications. An intriguing application in the construction industry is shown in Figure 4.2. Apart from this, a completely additively manufactured car (2010), food printing modeller (2011), first off-earth space printing machine purchase by NASA (2014), carbon clip technique (2015), concrete as well as bone printing (2017), 3D printed villa, etc, reflect the quantum of recent innovative AM applications.

In this chapter, a discussion on timely evolution and detailed timeline of AM technology has been given. The next chapter presents the details of a generalized AM process chain and its basic eight steps. AM as a part of

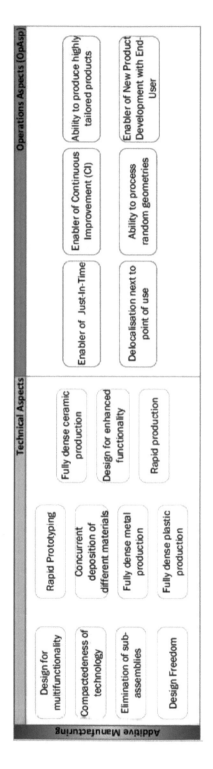

FIGURE 4.1

Opportunities in AM. (*Source:* [5]. Reprinted from Busachi et al., A review of additive manufacturing technology and cost estimation techniques for the defence sector. *CIRP Journal of Manufacturing Science and Technology,* 19: pp. 117–128, Copyright 2017, with permission from Elsevier.)

FIGURE 4.2
AM Printed Villa by Winsun. (*Source:* [6]. Reprinted from Wu et al., A critical review of the use of 3-D printing in the construction industry. *Automation in Construction*, 68: pp. 21–31, Copyright 2016, with permission from Elsevier.)

time compression engineering is introduced. Basic variations between various AM modellers; few maintenance and material handling issue will also be discussed in the forthcoming chapter.

References

1. ASTM F2792–12a, *Standard terminology for additive manufacturing technologies*, 2012.
2. Rathee, S., Srivastava, M., Maheshwari, S., Kundra, T. K., Siddiquee, A. N, *Friction based additive manufacturing technologies: principles for building in solid state, benefits, limitations, and applications*. 1st ed., 2018, CRC Press, Taylor & Francis Group.
3. Caffrey, T., Wohlers, T., Campbell, I., *Wohlers report 2017: 3D printing and additive manufacturing state of the industry annual worldwide progress report*, 2017, Wohlers Association Inc.

4. Attaran, M., The rise of 3-D printing: the advantages of additive manufacturing over traditional manufacturing. *Business Horizons*, 2017, **60**(5): pp. 677–688.

5. Busachi, A., Erkoyuncu, J., Colegrove, P., Martina, F., Watts, C., Drake, R., A review of additive manufacturing technology and cost estimation techniques for the defence sector. *CIRP Journal of Manufacturing Science and Technology*, 2017, **19**: pp. 117–128.

6. Wu, P., Wang, J., Wang, X., A critical review of the use of 3-D printing in the construction industry. *Automation in Construction*, 2016, **68**: pp. 21–31.

5

Generalized Additive Manufacturing Process Chain

5.1 Introduction

There are a few standard steps that need to be followed for generating useful physical parts via the AM route. The process chain basically refers to the standard steps that needs to be followed in order to completely accomplish the part fabrication objective. This chapter covers the generalized AM process chain. An important point to understand is that the process chain is dynamic in nature and is continuously evolving with development in current technologies, advent of newer ones and obsolescence of some existing practices. Also, the relevance of each step may have varied value to a specific process. While the same step may be essential for one AM technique, it may have trivial importance for another one. This chapter discusses the role of AM as an integral part of time compression engineering, presents discussion of data as well as information flow and details the eight basic AM process chain steps. It then proceeds to discuss the maintenance and material issues in AM.

5.2 AM as Fundamental Time Compression Engineering (TCE) Element

TCE, also known as concurrent engineering, has continuously evolved over the last three decades in the direction of integration of CAD, CAM, CAE, CAPP (computer aided design, manufacturing, engineering and process planning respectively) and RP/T/M (rapid prototyping/tooling/manufacturing). Three-dimensional CAD modelling is the key enabler behind TCE. The ability of concurrency in various operations in TCE imparts the freedom to considerably compress the conventional design cycle, as also the possibility of design iterations with minimal wastage. There is negligible duplication or miscommunication owing to utilization of centralized CAD data. AM is an integral part of TCE as depicted in Figure 5.1 [1]. Virtual prototyping (VP), which is an integral part of AM, is of utmost value in validating assemblies and reducing errors.

3D CAD models act as a gateway to RP/T/M and AM. Data exchange formats and representation techniques are especially important aspects of CAD for completely understanding inputs to AM processes. Therefore, these need to be thoroughly analyzed and understood. The three different modelling techniques, i.e. wireframe, surface and solid models, need thorough understanding for the effective design of three-dimensional models for various objects. Since each modelling software utilizes a different technique for representation of data, exchange between various systems becomes difficult. Two different approaches to data exchange between different systems are available: the first is the direct and the second is the indirect approach of data transfer. The direct approach involves creation of interface amid systems where data exchange is required. The indirect approach utilizes a neutral database for exchanging information which is non-vendor specific. AM uses the second approach, i.e. the indirect one. Many neutral formats were initially suggested, including:

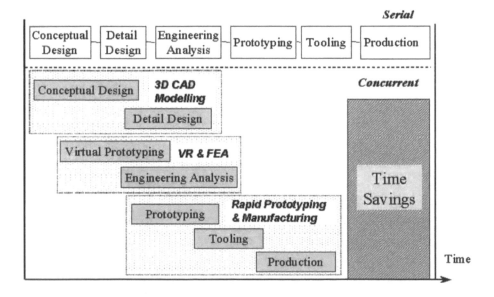

FIGURE 5.1
Rapid Prototyping/Manufacturing/AM as a Part of TCE. (*Source:* [1]. Reprinted from Pham and Dimov, Introduction, in *Rapid manufacturing: the technologies and applications of rapid prototyping and rapid tooling*, 2001, pp. 1–18, Springer London, with permission from Springer Nature.)

1. STL (STereoLithography)
2. SLC (3D System SLice Contour)
3. RPI (Rapid Prototyping Interface)
4. LMI (Layer Manufacturing Interface)
5. CLI (Common Layer Interface)
6. IGES (Initial Graphical Exchange Specification)
7. HPGL (Hewlett-Packard Graphics Language)
8. LEAF (Layer Exchange Ascii Format)
9. STEP (STandard for Exchange of Product model data)
10. VRML (Virtual Reality Modelling Language).

However, today STL is considered the de facto standard to interface CAD and AM systems. The remaining formats have been proposed for addressing STL shortcomings but in fact have restricted utility. Accurate CAD models are tessellated to obtain .stl files and their surfaces are approximated as triangles where each triangle is defined by four parameters (three vertex and one outward normal). ASCII and binary (preferable owing to lesser size) formats are used to store STL files.

5.3 AM Data and Information Flow

Flow of information and data is common in all AM systems and consists of a few basic steps. A list of these steps (in order) is summarized below:

- Creation
- Export
- Validation and export
- Orientation and scaling
- Generating support structures
- Adjusting process parameters
- Slice data creation.

Figure 5.2 and 5.3 present the data and information flow in the AM processes. 3D data is created by some suitable technique as per availability. This data is exported with the aid of a neutral file format. Most advanced AM packages have a validation and repair feature which ensures aptness of exported data. The part is then oriented in the build volume to compensate for probable anomalies. Support structures are then generated as per

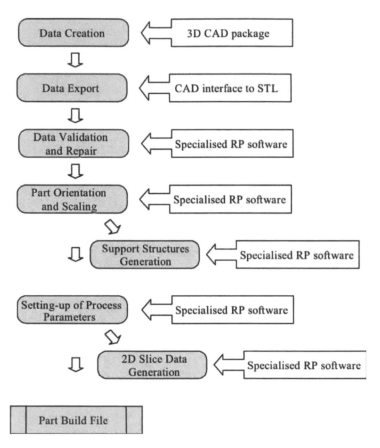

FIGURE 5.2
Flow of Information in AM Processes. (*Source:* [1]. Reprinted from Pham and Dimov, Introduction, in *Rapid manufacturing: the technologies and applications of rapid prototyping and rapid tooling*, 2001, pp. 1–18, Springer London, with permission from Springer Nature.)

requirement. Build styles, tooling and other system parameters are then adjusted to suit requirements. Finally, slicing is accomplished for actual fabrication of layered artefacts.

5.4 Generalized AM Process Chain and Eight Steps in AM

In most commercial AM processes, component fabrication is accomplished by layered material addition where every layer has very constrained thickness and is a component section obtained from 3D CAD data. Since

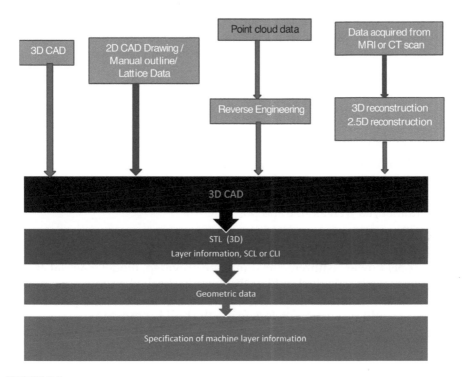

FIGURE 5.3
Data Flow in AM Processes. (*Source:* [2]. Reproduced from Srivastava, *Some studies on layout of generative manufacturing processes for functional components*, 2015, Delhi University, India.)

infinitesimally small thicknesses are not feasible, the final component approximates to the original data owing to finitely thick layers. It should be noted that the accuracy of parts is directly proportional to the number and inversely proportional to thickness of each slice [2].

To accomplish this, each AM process follows a series of steps in the process of conversion of 3D CAD data to a physical part. There is a variation in the extent and technique of AM utilization with change in part type. While simple parts can be directly printed, complicated parts need iterative designs and a multitude of considerations. Also, concept models and prototypes can be roughly prepared but components to be used as end products require careful planning as well as post-processing.

Eight fundamental AM steps that are more or less common to each technique can be understood as follows:

Step 1: Obtaining 3D CAD data of the desired part
    3D CAD data which fully describes the part to be fabricated is obtained either directly from professional 3D software or from 2D

data with additional information, or from reverse engineering or any other means. An important point to be kept in mind is that this data should fully represent the solid/surface for the part.

Step 2: Conversion of data from step 1 into .stl format
Most of the AM modellers take input in the form of .stl format which can be obtained directly as input from the CAD software utilized in step 1. External closed surfaces are described in this format and slices are calculated based upon this format.

Step 3: Transferring .stl file to modeller
This .stl file is transferred to the AM modeller along with necessary inputs regarding the tool path generation, machine parameters, orientation, as well as size and position.

Step 4: Setting up machine parameters
The machine should be properly set up, pre-warmed if necessary, proper power requirements should be provided, model material and raw material spools should be thoroughly checked, proper operation of valves, compressor, machine build center panel, build platform, etc. should be ensured before firing the build command.

Step 5: Part building
The superiority of AM processes emanates from the fact that the part building process is completely automated. However, intermittent checking can be undertaken to ensure power consistency, spool malfunction, software related issues, etc.

Step 6: Part removal from the modeller
This has to be undertaken in a judicious manner once the part is finished. This may require consideration of temperature of build volume, some inherent interlocking features of the machine, support layers at base, etc.

Step 7: Post-processing
There is a need to clean up the part before it can be put to final use. Some support features may be required to be taken care of with due consideration of the strength of the parts. Wherever required, this demands skill and expertise.

Step 8: Applications
Before final application, treatments like painting, improving texture, priming, improving surface finish and so on may be required. Sometimes, assembling individual components is also required if the parts are sub-components of a larger part.

However, for the ease of understanding, some of these eight steps can be combined and reclassified as the five AM steps named below and presented in Figure 5.4:

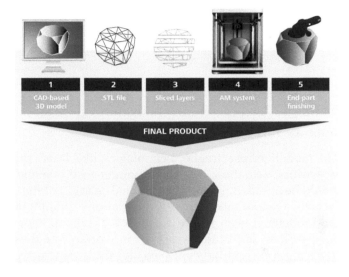

FIGURE 5.4
AM Process Flow Chain. (*Source:* [3]. Cotteleer et al., *The 3D opportunity primer: the basics of additive manufacturing*, 2014, Deloitte University Press, Westlake, Texas. Available from: http://dupress.com/articles/the-3d-opportunity-primer-the-basics-of-additive-manufacturing, accessed February 8, 2019, with permission.)

- CAD-based modelling
- Preparation of .stl file
- Slicing of layers
- Actual part building
- Finishing of end part.

5.5 Variation from One AM Machine to Another

An important point to remember is that the basic underlying principle in nearly all commercial AM modellers is layer-based. However, the main differences between AM machines consist in the basis of raw materials used, creation of layers and mechanism of bonding amongst layers. These factors appreciably affect the accuracy, material and mechanical characteristics of the fabricated parts. Also, speed of fabrication, overall build time, magnitude of post-processing required, modeller dimensions, total cost of modeller, as well as AM processes are affected by these parameters. Thus, there are multiple ways to classify AM processes, as has already been covered in Chapter 3. Also, a brief overview of various AM machines has already been presented in Chapter 1 of this book.

5.6 Maintenance of Equipment

Since the AM modellers use fragile technologies like laser, printer, computer, wires, sensors, actuators, etc., their maintenance is mandatory. Dirt and noise should be avoided. Frequent locational changes of the AM facility should be strictly discouraged. Proper foundations should be used for mounting the modellers; even where frequency of jolts is lower misalignment can have serious implications upon the working of these systems. These systems should be regularly checked and the maintenance schedules must be strictly followed.

An important fact here is that standardization of AM tests and materials has still not been fully accomplished and limited standards are available for AM systems. ASTM F42 technical committee is working seriously in this direction. However, in most cases the manufacturer's manual contains standard testing procedures corresponding to specific modellers to ensure their smooth functioning. Further discussion of various associated technologies is covered in Section C of this book.

5.7 Material Handling Issues

Apart from the modeller, careful handling of AM materials is also a prerequisite since they possess limited shelf lives and tendency to undergo chemical reactions. Care should be taken to avoid moisture, light exposure, contamination, etc. Though recycling of materials is feasible in many cases, yet it is necessary to ensure that the material maintains consistency of properties during repeated work cycles.

5.8 Summary

AM has prominently established itself as a fundamental TCE element owing to its capabilities of considerably compacting the product design and development process. The role of 3D modelling as a gateway to AM is explained. Also, the pros and cons of present geometrical representation methods are highlighted. Description of formats to interface CAD with AM is subsequently followed by principal steps description to generate requisite data for part fabrication in layers. To stand out as an AM designer, one should keep in mind the following:

1. Maintain versatility and take modelling as a challenge.
2. Learn basics of some programming language for sculpting, like ZBrush or Mudbox.
3. Learn texturing skills.
4. Learn basic anatomy of objects that you need to model.

This chapter discusses the role of AM as an integral part of TCE. It then proceeds to discuss the data and information flow in AM. The AM process chain comprising of eight general steps is then discussed in detail. Variation of characteristics with varying modellers is then highlighted along with a brief summary of maintenance and material handling issues of AM modellers. This chapter ends with a concluding summary. This chapter concludes the first section of this book. This section covered the various general aspects of AM. The next section covers the process specific details of various AM processes.

References

1. Pham, D. T., Dimov, S. S., Introduction, in *Rapid manufacturing: the technologies and applications of rapid prototyping and rapid tooling*, 2001, Springer, pp. 1–18.
2. Srivastava, M., *Some studies on layout of generative manufacturing processes for functional components*, 2015, Delhi University, India.
3. Cotteleer, M., Holdowsky, J., Mahto, M., *The 3D opportunity primer: the basics of additive manufacturing*, 2014, Deloitte University Press, Westlake, Texas. Available from: http://dupress.com/articles/the-3d-opportunity-primer-the-basics-of-additive-manufacturing, accessed February 8, 2019.

Section B

Process Specific Details of Various Additive Manufacturing Processes

6

Additive Manufacturing Processes Utilizing Vat Photopolymerization

6.1 Introduction

Different vat photopolymerization processes are introduced in this chapter. Stereolithography is one of the oldest and best-known examples of such processes and is popular to the extent of it being used synonymously with the vat polymerization process itself. ISO/ASTM defines this process as: "Vat photopolymerization is an additive manufacturing process in which liquid photopolymer in a vat is selectively cured by light-activated polymerization" [1]. Dr. Hideo Kodama (1981) was the pioneer in the stereolithography (SLA) process as a fast and economical substitute for holographic techniques. Dr. Charles Hull (1986) filed the first patent on SLA and founded 3D Systems Inc. SLA was first used for plastic part creation. However, with the passage of time this technique has proven its versatility in fabrication of a variety of intricate and customized parts. The SLA process is capable of making parts from metals, ceramics, plastics, composites and so on.

This is the first chapter of Section B of this book, which presents process specific details of various AM processes and consists of seven chapters. This chapter provides details of AM processes using vat photopolymerization in terms of various material related aspects including: precursors, photoinitiators, absorbers, filled resins, additives and postprocessing; details of the photopolymerization process; process modelling aspects; variants and classification of the vat photopolymerization process, including free and constrained surface approach, Laser-SLA, digital light processing SLA process (DLP-SLA), liquid crystal display stereolithography (LCD-SLA); and advantages and limitations of vat photopolymerization processes. It concludes the discussion with a summary.

6.2 Materials

Vat photopolymerization processes use plastics as well as polymeric materials, especially UV-curable photopolymeric resins. An example of resins used with these processes is the Visijet range (3D systems). There are various integral parts of any photocurable SLA resin which include precursor, photo initiator, additives, absorber and filler as shown in Figure 6.1. These are explained in detail in the subsequent sections.

6.2.1 Precursors

These are liquid molecules that link mutually or polymerize when exposed to light for obtaining solid three-dimensional network. Monomers, oligomers, prepolymers, etc. can be utilized based upon specific requirements. Acrylate-based resins which are based upon radical photopolymerization possess high reactivities and are available in different types based upon the

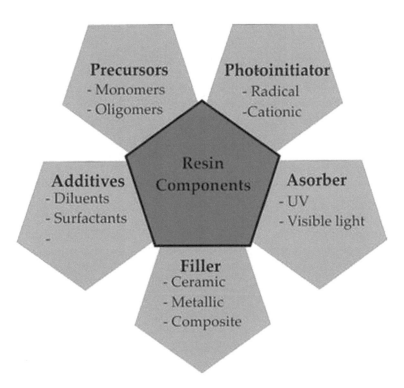

FIGURE 6.1
Resin Components in SLA. (*Source:* [2]. Schmidleithner and Kalaskar, Stereolithography, in *3D printing*, D. Cvetković, Editor, 2018, IntechOpen, under Creative Commons Licence.)

number of reactive groups or oligomer types for specific requirements of mechanical and thermal resistances. They offer the basic advantage of fast build speeds but exhibit high shrinkage during the printing process, which is their major limitation since distortion of parts occur. Another disadvantage is the oxygen sensitivity of acrylate-based resins due to which the polymerization reaction suffers a setback. The next important class of resins is epoxy resins which are based upon cationic reactions that offer reduced sensitivity to oxygen and shrinkage compared with acrylates. They are however sensitive to moisture. Hybrid systems are also in use utilizing the favorable features of individual precursors. For example, a combination of acrylate and epoxy resins is very commonly used and results in low shrinkage materials displaying fast curing rates.

6.2.2 Photoinitiators

Photoinitiators (PIs) are reactive to light. Irradiation of PIs with lights of the right wavelength leads to their excitation and thus initiation of the curing reaction. Correct choice of PI is therefore important and also depends on precursor type. It also effects a number of important SLA characteristics including kinetics, cross-linking density, mechanical characteristics, etc. to name a few.

6.2.3 Absorbers

Absorbers limit light penetration beyond the desired cure depths. Benzotriazole derivative is a common UV absorber. A precise control is very important for intricate geometries with overcuts otherwise feature loss will result.

6.2.4 Filled Resins

Resins are filled with powders to fabricate metal/ceramic parts. After filling resins with powders, the parts are printed using the standard vat photopolymerization technique, debinded for removal of organic resin parts via pyrolysis and sintered using heat treatment to obtain final dense parts. This process is shown in Figure 6.2. Shrinkage coefficients of different metals need to be taken into account for specific geometrical requirements. To minimize shrinkage, high filler amounts are required. It should be kept in mind that size of particle should be smaller than layer height. If the size of particles approaches the wavelength of incident light, then scattering phenomenon becomes important and needs to be effectively dealt with in obtaining final cure depth and accuracy. The difference in refractive index between filler and matrix should be minimal to counteract scattering. If nano-sized particles are added, then the properties of parts obtained can be further improved.

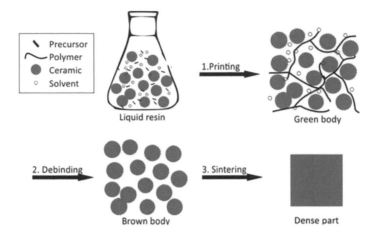

FIGURE 6.2
Typical Steps in Fabrication of Dense SLA Ceramic Parts Using Filled Resins. (*Source:* [2]. Schmidleithner and Kalaskar, Stereolithography, in *3D printing*, D. Cvetković, Editor, 2018, IntechOpen, under Creative Commons Licence.)

6.2.5 Additives

Many additives are added to counteract the negative effects of high-volume fraction of solid loading that can change resin flow behavior, affect coating mechanism, increase requirement of mechanical forces for lifting platform and so on. Rheological additives as well as stabilizers are added which can enhance solid loading, their shelf life as well as stability. The particles agglomerate and sediment during the process, which can be prevented by using oligomeric surfactants, oleic acid (long chain acid) or phosphine oxides to obtain homogeneity in ceramic/metal powder distribution.

6.2.6 Post-Processing

When the part is removed from the build platform, support structures are eliminated from it. This involves cleaning in solvents, drying of structures and then sanding. The parts are then cured in a UV chamber to obtain enhanced mechanical characteristics. Debinding and sintering are considered as post-processing steps in the case of filled resins.

6.3 Photopolymerization Process

These systems utilize a vat of liquid photopolymer resin. Layers are obtained by curing or hardening this resin using ultraviolet (UV light) adjusted through motor-controlled mirrors according to a predefined scan strategy. A photo initiator (PI) molecule is responsive to incident light in a fashion that it initiates a localized chemical photopolymerization reaction and hence curing. The platform holding the vat lowers each time a new layer is required to be cured. This process of layered artefact creation is common to all SLA processes. Support structures are specially required in these systems since there is no support from adjacent material during part fabrication owing to liquid state raw material. Figure 6.3 shows the main schematic of the vat photopolymerization process and Figure 6.4 presents the generalized process chain of the vat photopolymerization process.

The main steps of the vat photopolymerization process are defined below:

1. Lowering of build platform from resin vat top by one-layer thickness
2. Curing of resin by UV light and subsequent lowering of platform to accommodate next layer
3. Smoothening of resin to provide base for next layer
4. Draining of resin from vat
5. Removal of part
6. Removal of support
7. Post-curing.

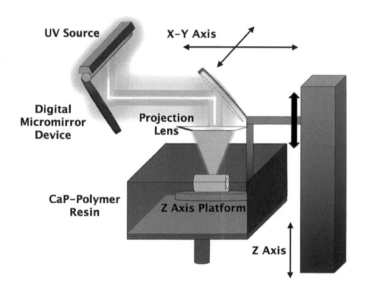

FIGURE 6.3

Schematic of Vat Photopolymerization. (*Source:* [3]. Reprinted from Trombetta et al., 3D printing of calcium phosphate ceramics for bone tissue engineering and drug delivery. *Annals of Biomedical Engineering*, 2017, 45(1): pp. 23–44.)

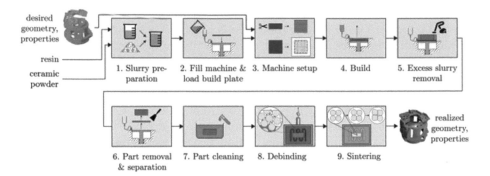

FIGURE 6.4
Vat Photopolymerization Process Chain (for Ceramic). (*Source:* [4]. Reprinted from Hafkamp et al., A feasibility study on process monitoring and control in vat photopolymerization of ceramics. *Mechatronics*, 2018, 56: pp. 220–241, with permission from Elsevier.)

The last three steps of removal of part, removal of support and post-curing together constitute the post-processing stage. It involves removal of parts from resin and completely draining of any excess resin from the vat. A knife or sharp implements are utilized to remove supports. Appropriate safety precautions should be taken to prevent contamination of resin. There are many ways to remove resin and supports. One of these is rinsing with alcohol and then with water. Parts are either allowed to dry naturally or by using an air hose. UV light is utilized to finally post-cure parts to obtain good quality parts.

The process and use of support often results in defects like air gaps that should be filled with resin for achieving high-quality models. Layer thickness obtained is typically around 0.025–0.5 mm.

6.4 Process Modelling

In any process, it is imperative to develop a clear understanding of the physics-based mathematical process models that are involved and which result in useful transformation of input into output. This section aims to introduce the readers to the various process models involved for the photopolymerization process. A self-explanatory schematic to illustrate the process is presented in Figure 6.5 which shows the relationship between input, output and system parameters for a photopolymerization process by breaking it into two sub-models.

There are various parameters that are important in the modelling of irradiation phenomenon. These include irradiation profiles, the way in which beam

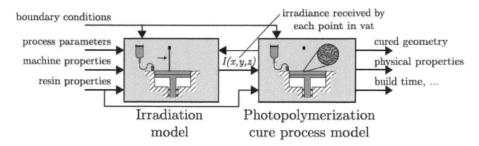

boundary conditions

process parameters

machine properties

resin properties

$I(x,y,z)$

irradiance received by
each point in vat

cured geometry

physical properties

build time, ...

Irradiation
model

Photopolymerization
cure process model

FIGURE 6.5

Block Diagram for Process Modelling of Photopolymerization Model. (*Source:* [4]. Reprinted from Hafkamp et al., A feasibility study on process monitoring and control in vat photopolymerization of ceramics. *Mechatronics*, 2018, 56: pp. 220–241, with permission from Elsevier.)

is propagated, attenuation and absorption, scattering, refraction and optical self-focussing and beam positioning. Again, there are many important terms in the modelling of the photopolymerization cure process phenomenon, including exposure and its threshold model, photopolymerization cure process phenomenon, degree of conversion, oxygen inhibition, heat transfer, mass transfer, shrinkage and distortion and mechanical property variation. Many modelling techniques are available to model all these individual items. A non-exhaustive list of comparison of these is presented in Table 6.1.

6.5 Variants and Classification of Vat Photopolymerization Process

Various vat photopolymerization processes include continuous liquid interface production (CLIP), scan, spin and selectively photocure technology (3SP), solid ground curing (SGC), stereolithography (SL), stereolithography apparatus (SLA), two-photon polymerization (2PP), etc. However, owing to SLA processes being pioneers in their category, vat photopolymerization techniques are used synonymously with SLA processes.

SLA processes can be classified on two main bases as shown in Figure 6.6. The first basis is the irradiation technique. If the laser scans each cross-sectional point of a given area, then the process is termed laser SLA. If the pixelated image is projected in one go, then the method is called digital light processing (DLP). If a LCD photomask is utilized for irradiation then the process is called LCD-SLA.

Another common classification is based upon the incident light direction. In the first method, light incidence is accomplished from above the top

TABLE 6.1

Comparison of Different Process Models for Photopolymerization Techniques

Process Model	Ceramic loaded?	Irradiation method	Irradiation profile	Beam propagation	Attenuation	Scattering	Refraction	Beam motion
Ceramic µSL model	Yes	Vector scanning	Gaussian; Monte Carlo ray tracing	Yes	Time-varying Beer–Lambert	Mie theory	No	Stationary
Effective Beer's law model	Yes	Vector scanning	Gaussian	No	Beer–Lambert	Yes	No	Scanning at fixed speed
Predictive ceramic photopolymerization model	Yes	Vector scanning	Gaussian	No	Beer–Lambert	Yes	No	Scanning at fixed speed
General photocurable model of resin with microparticles	Yes	Vector scanning	Gaussian	No	Beer–Lambert	Yes	No	Scanning at fixed speed
Exposure threshold model	No	Vector scanning	Gaussian	No	Beer–Lambert	No	No	Scanning at fixed speed
Analytical irradiance model	No	Vector scanning	Gaussian	Yes	Beer–Lambert	No	At vat surface	Scanning
Non-isothermal photo-polymerization model	No	Vector scanning	Gaussian	No	Time-varying Beer–Lambert	No	No	Stationary and scanning
Stereolithography simulation model	No	Vector scanning	Gaussian	No	Time-varying Beer–Lambert	No	No	Stationary
SLA cure model	No	Vector scanning	Gaussian and top-hat	No	Time-varying Beer–Lambert	No	No	Scanning at fixed speed
Deterministic photopolymerization model	No	Vector scanning	Gaussian		Beer–Lambert	No	No	Scanning at fixed speed
Space-resolved photo-polymerization model	No	Vector scanning	Gaussian with noise	No	Time-varying Beer–Lambert	No	At cure front	Stationary
Diffusion-limited photopolymerization model	No	Vector scanning	Gaussian	No	Time-varying Beer–Lambert	No	No	Stationary

Continued

Model								
Mathematical model for photopolymerization	No	Vector scanning	Gaussian	No	Beer–Lambert	No	No	Stationary
Photo-thermal-kinetic model	No	Vector scanning	Gaussian	No	Beer–Lambert	No	No	Stationary
STLG-FEM	No	Vector scanning	Gaussian	No	Beer–Lambert	No	No	Stationary
Phenomenological FEM model	No	Vector scanning	Gaussian	No	Beer–Lambert	No	No	Scanning at fixed speed
Thermal deformation model	No	Vector scanning	Gaussian	No	Time-varying Beer–Lambert	No	No	Scanning at fixed speed
Dynamic FEM model	No	Vector scanning and mask projection	Gaussian	No	Time-varying Beer–Lambert	No	No	Scanning at fixed speed
Transient layer cure model	No	Mask projection	Exact ray traced pixels	Yes	Time-varying Beer–Lambert	No	No	No (varying bitmaps)
Semi-empirical material model	No	Mask projection	Exact ray traced pixels	Yes	Beer–Lambert	No	No	Stationary
Pixel-based solidification model	No	Mask projection	Non-axisymmetric Gaussian	No	Beer–Lambert	No	No	Stationary
Analytical model for scanning projection based SL	No	Scanning mask Projection	Gaussian	No	Beer–Lambert	No	No	Scanning at fixed speed
Comprehensive free radical photopolymerization model	No	NA	1D	No	Time-varying Beer–Lambert	No	No	Stationary
Thermal model of stepless rapid prototyping process	No	NA	1D	No	Time-varying Beer–Lambert	No	No	Stationary
1D UV curing process model	No; GFRP	NA	1D	No	Beer–Lambert	No	No	Stationary

TABLE 6.1 *Continued*

Process Model	Curing mechanism	DOC	Oxygen inhib.	Heat transfer	Mass transfer	Shape formation	Shrinkage	Mechanical property variation	Spatial dim.
Ceramic µSL model	Free radical	Yes	No	No	No	Yes	No	No	2D
Effective Beer's law model	None	No	No	No	No	Exposure threshold	No	No	2D
Predictive ceramic photopolymerization model	None	No	No	No	No	Exposure threshold	No	No	2D
General photocurable model of resin with microparticles	None	No	No	No	No	Exposure threshold	No	No	
Exposure threshold model	None	No	No	No	No	Exposure threshold	No	No	2D
Analytical irradiance model	None	No	No	No	No	Exposure Threshold	No	No	3D
Non-isothermal photopolymerization model	Free radical	Yes	No	Polymerization heat	No	Yes	Polymerization	No	2D/3D
Stereolithography simulation Model	Free radical	Yes	No	Polymerization heat	No	No	No	No	3D
SLA cure model	Free radical	Yes	No	Polymerization heat+convection	Monomer and radical diffusion	Conversion threshold	No	No	2D
Deterministic Photopolymerization model	Free radical	Yes	Yes	Polymerization heat	Monomer, radical and oxygen diffusion	Conversion threshold	No	No	2D
Space-resolved photopolymerization model	Free radical	No	No	No	No	Yes	No	No	2D

Model									
Diffusion-limited photopolymerization model	Free radical	Yes	No	Polymerization heat	No	No	No	No	2D
Mathematical model for photopolymerization	Free radical	Yes	No	No	No	Yes	No	No	2D
Photo-thermal-kinetic model	Free radical	Yes	No	Polymerization heat	No	No	No	No	2D
STILG-FEM	Free radical + thermal	Yes	No	Polymerization heat	No	No	Polymerization	No	2D
Phenomenological FEM model	None	No	No	No	No	Element birth	Polymerization	No	3D
Thermal deformation model	None	Yes	No	Polymerization heat	No	No	No	No	2D
Dynamic FEM model	None	No	No	Polymerization heat	No	No	Polymerization + thermal	E-modulus	3D
Transient layer cure model	None	No	Not explicit	No	Not explicit	Exposure threshold	No	No	3D
Semi-empirical material model	None	No	No	No	No	Exposure threshold	No	No	2D
Pixel based solidification model	None	No	No	No	No	Exposure threshold	No	No	3D
Analytical model for scanning projection-based SL	None	No	No	No	No	Exposure threshold	No	No	2D
Comprehensive free radical photopolymerization model	Free radical	Yes	Yes	Polymerization heat + laser absorption	Diffusion of all species	No	No	No	1D
Thermal model of stepless rapid prototyping process	Free radical	No	No	Polymerization heat + laser absorption	No	No	No	No	1D
1D UV curing process model	Free radical	Yes	No	Polymerization heat	No	No	No	No	1D

Source: [4]. Hafkamp et al., A feasibility study on process monitoring and control in vat photopolymerization of ceramics. *Mechatronics*, 2018, 56: pp. 220–241.

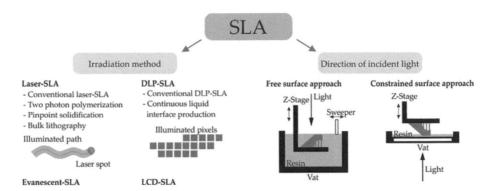

FIGURE 6.6
Basis of Classification of SLA Process. (*Source:* [2]. Schmidleithner and Kalaskar, Stereolithography, in *3D printing*, D. Cvetković, Editor, 2018, IntechOpen, under Creative Commons Licence.)

surface. This is called the free surface approach. In the second method, light incidence is performed from the bottom of a transparent vat. This is called the constrained surface approach.

6.5.1 Free and Constrained Surface Approaches

In the free surface approach, the build platform is located in the resin tank and is coated with it, amounting to the growth of first part layer. The film is cured from above the resin bath. Lowering of platform takes place at the completion of the first layer and similarly the next layer is formed after coating of resin using a mechanical sweeper.

The constrained surface approach is also called the bottom exposure approach. Here, there is a suspended build platform above a resin bath. Illumination is accomplished through a transparent floor. This leads to curing of the resin layer between build platform and vat floor. The platform is then raised by a predefined level. The final part is suspended upside down. Supports made of the same printing material are used for overhangs and undercuts to ensure sufficient adhesive forces.

The second technique, i.e. bottom exposure approach, is more common owing to several advantages over the former, i.e. free surface approach. One of these is the smooth surfaces obtained via this route. Another advantage is absence of mechanical sweeper and less printing time. Also, less resin is used in this technique. However, this approach has its own limitations, which include a need to overcome the attractive force between vat floor and each layer of the part to be made. Sometimes a hydraulic layer is applied to counteract this interaction or by modifying mechanical separation mechanisms or by application of shear forces.

6.5.2 Laser-SLA

This technique is called laser SLA, vector-based SLA or SLA. Each resin film layer is cured by scanning UV light onto it. Laser movement in the x–y plane is accomplished by using two galvanometers and a dedicated optical system. These are slightly more expensive than their contemporaries but offer higher resolution (5–10 microns) with careful choice of parameters, including z-axis accuracy, resin composition, mechanism of UV scanning, geometry of laser lines, optical system characteristics, etc. There are three methods for further improvement of resolution to sub-micro level which are defined below.

6.5.2.1 Two Photon Photopolymerization (TPP)

TPP is used to obtain resolutions around 100 nanometres and roughness below 10 nanometres. It was first introduced by Denk et al. [5] and was first commercialized by Nanoscribe GmbH [6]. Here the photo initiator excitation and thus curing does not occur in the total path that is illuminated by laser. It occurs only in its focal point region, termed volume pixel/voxel. Simultaneous absorption of two photons occurs due to highly intense femtosecond pulsed laser. Since the probability of this process is dependent on laser pulse intensity, laser focal point becomes a critical factor. Owing to higher wavelengths and corresponding less energy, lasers such as titanium-sapphire-based ones are used instead of normal UV lasers. Excitation energy is obtained from the combined energy of individual photons. 3D voxels in TPP enable resin cure inside the bath, also thereby easily eliminating the requirement for support structures and freely moving components. However, there is a big restraint upon the size of parts produced by this method (millimetre size). Also, the reduced speed of laser lines is an issue. A lot of research has been directed towards attempts at improving the size of components achieved by TPP. A typical set-up for a TPP system is presented in Figure 6.7.

6.5.2.2 Pinpoint Solidification

The pinpoint solidification method involving single photon photopolymerization is also known as super integrated hardened polymer stereolithography (Super IH SLA) and was first suggested by Ikuta et al. [8]. Due to focussed high intensity of the laser at its focal point, resin is cured at a voxel. It is possible to obtain very high accuracy (0.4 microns) without need of an expensive laser. This process is still not commercial and there is limited research into this SLA variant.

6.5.2.3 Bulk Lithography

In bulk lithography, three-dimensional texturing is achieved by varying exposure energy. Depth of features is defined by the cure depth which

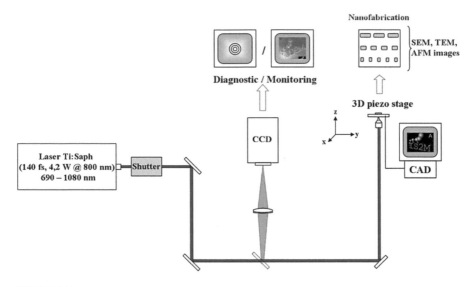

FIGURE 6.7
Schematic for Typical Experimental Set-Up for TPP. (*Source:* [7]. Spangenberg et al., Recent advances in two-photon stereolithography, Chapter 2, in *Updates in advanced lithography*, 2013, InTechOpen, pp. 35–62., under Creative Commons Licence.)

enables the part to be one layer of varying thickness, thus totally eliminating the stair-stepping defect common to most AM techniques. Overhanging features are not possible and only thin parts can be obtained via this technique.

6.5.3 Digital Light Processing SLA Process (DLP-SLA)

In DLP, simultaneous illumination of the entire cross-sectional area is accomplished with the aid of a DLP light engine. A digital micrometer device (DMD), which consists of an array of mirrors, is a critical element and serves as a dynamic mask during DLP. Every mirror is individually tilted to quickly and reliably switch pixels. The DMD is linked to a computer, light source and an optical set-up, projection of desired light cross-sections is accomplished in a quick and precise fashion. A real-time image of this printer and a schematic diagram to show its working principle along with printed object cross-section is presented in Figure 6.8. Grayscale illumination can be obtained owing to the fast DMD switching rate. Exposure time as well as energy can therefore be precisely controlled. Resolutions of around 25 microns can be obtained by this process, including least feature size of 0.6 microns and ceramic filled resins with linear and lateral dimension as 15 and 40 microns respectively.

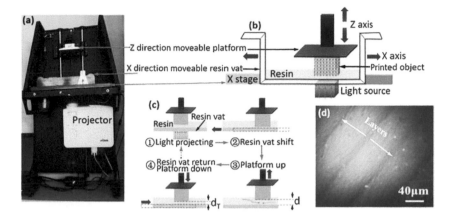

FIGURE 6.8
DLP-SLA Printer: (a) Real Time DLP B9Creator 3D Printer (without Sweeper); (b) Schematic Illustration of Working; (c) Steps in Working; (d) Cross-Section of Printed Part. (*Source:* [9]. Reprinted from Mu et al., Digital light processing 3D printing of conductive complex structures. *Additive Manufacturing*, 2017, 18: pp. 74–83, Copyright 2017) with permission from Elsevier.)

This technique can be utilized to make a variety of parts utilizing raw materials like multi-wall carbon nanotubes (MWCNT) nanocomposites and pristine resins which include springs, resistors, truss, capacitors, capacitor arrays, capacitors at variable heights, etc. in different colors. A few of these are presented in Figure 6.9.

Since the principle of pixels is involved in DLP processes, corners are sharp but curved surfaces obtained via this method are saw-toothed rough. This process has a limitation on the resolution and is dependent on pixel size, optical system, number of mirrors, etc. Also, printing of large parts is usually unsatisfactory in terms of accuracy as compared to smaller parts. However, the printing speeds are reasonably good and economical in case of DLP systems.

6.5.3.1 Continuous Liquid Interface Production

The continuous liquid interface production process (CLIP) was commercialized by Carbon Inc. It is characterized by the creation of a dead interface zone since the resin between build platform and material tray is uncured and retains its liquid state. A vat floor possessing oxygen permeability is utilized for the generation of this zone. Such an arrangement prevents curing. There is no need for a recoating mechanism. Build platform can be continuously elevated thereby reducing surface roughness. The printing rates speed up to the order of 500mm/hour. Liquids like brine that are highly dense, inert and do not mix with the liquid layer are utilized for creation of the interface film.

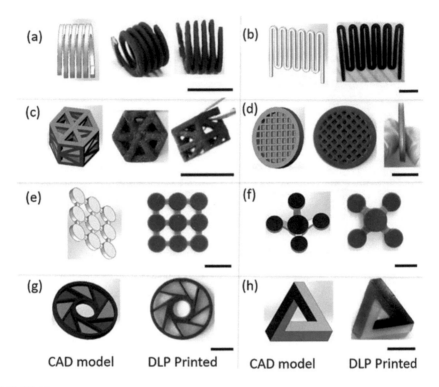

FIGURE 6.9
Different Components Made by DLP SLA Route, Scale Bar of Parts is Kept at 10 millimetres. (*Source:* [9]. Reprinted from Mu et al., Digital light processing 3D printing of conductive complex structures. *Additive Manufacturing*, 2017, 18: pp. 74–83, Copyright 2017, with permission from Elsevier.)

6.5.4 Liquid Crystal Display Stereolithography

In the liquid crystal display stereolithography (LCD SLA) process, a liquid crystal display device utilizes a dynamic mask to carry out the stereolithography process. However, DLP SLA has replaced LCD SLA in most applications owing to their ability to switch speeds and their high accuracy. However, these systems are mainly economical substitutes to DLP methods.

6.6 Advantages and Limitations of Vat Photopolymerization Processes

Processes based upon vat photopolymerization possess a variety of advantages. These mainly include:

- High accuracy
- Good surface finish
- Process can be accomplished with single laser and optical system
- Defect free layers are obtained by the use of recoating blades
- Relatively quick
- Typically, large build volumes for supporting heavy objects and so on.

Despite many advantages, these processes are not free from limitations. A few major limitations are summarized below:

- Requirement for support structures
- Requirement for post-curing
- Use of recoating blades tends to increase the process cost
- Lengthy post-processing time due to need for scrubbing to achieve complete material removal
- Limited raw materials can be used.

6.7 Summary

Vat photopolymerization is a versatile technology with varied applications. The spectrum of functional parts obtained from this route is appreciably broad. High performance material SLA parts find multiple applications in the automobile and aircraft sectors. Silicones possessing high elasticity find application in soft robotics. Epoxies possessing considerable strengths that undergo post-curing find varied structural applications. This process offers excellent accuracy and surface properties to its parts. DLP-SLA is also a competitive technology for creating microfluidic devices and can offer excellent resolution, small and delicate features, good surface finish and appreciably high production rates. Filled resins can offer slightly better thermo-mechanical stability. Apart from these, many applications using vat photopolymerization-based techniques have been reported in the field of medicine, for example, craniofacial implants from hydroxyapatite; in the field of dentistry, for example, creation of crowns; as well as hearing aids, prostheses, drug delivery systems like microneedles, AM printed custom-ized tablets, medical imaging, tissue engineering, regenerative medicines, bioprinting.

However, the printing speeds are appreciably low and cost is almost the highest for parts fabricated via this route. Also, the SLA parts are not thermo-mechanically very stable. Handling of resins is an extremely cumbersome and messy process. This has led to a setback for this process owing to which it barely competes with other contemporary AM techniques. Though soft lithography is not actually an AM process, yet it is sometimes treated as an effective SLA substitute owing to its ability to provide better surface finish than most SLA systems. It is utilized as an alternative to SLA specially in microfluidic systems where extremely fine holes and cavities need to be drilled for applications like inkjet printheads.

In this chapter, an attempt has been made to cover all the necessary aspects of vat polymerization process. The next chapter covers the powder bed fusion process.

References

1. ASTM F2792–12a, *Standard terminology for additive manufacturing technologies*, 2012.
2. Schmidleithner, C., Kalaskar, D. M., Stereolithography, in *3D printing*, D. Cvetković, Editor, 2018, IntechOpen.
3. Trombetta, R., Inzana, J. A., Schwarz, E. M., Kates, S. L., Awad, H. A., 3D Printing of calcium phosphate ceramics for bone tissue engineering and drug delivery. *Annals of Biomedical Engineering*, 2017, **45**(1): pp. 23–44.
4. Hafkamp, T., van Baars, G., de Jager, B., Etman, P., A feasibility study on process monitoring and control in vat photopolymerization of ceramics. *Mechatronics*, 2018, **56**: pp. 220–241.
5. Denk, W., Strickler, J. H., Webb, W. W., Two-photon laser scanning fluorescence microscopy. *Science*, 1990, **248**(4951): pp. 73–76.
6. *Das Unternehmen im Detail – Nanoscribe GmbH*. Available from: www.nanoscribe.de/de/unternehmen, accessed March 10, 2018.
7. Spangenberg, A., Hobeika, N., Stehlin, F., Malval, J. P., Wieder, F., Prabhakaran, P., Baldeck, P., Soppera, O., Recent advances in two-photon stereolithography, Chapter 2, in *Updates in advanced lithography*, 2013, InTechOpen, pp. 35–62.
8. Ikuta, K., Maruo, S., Kojima, S. New micro stereo lithography for freely movable 3D micro structure-super IH process with submicron resolution, in *International Conference on Micro Electro Mechanical Systems (MEMS)*, 1998, IEEE, Heidelberg.
9. Mu, Q., Wang, L., Dunn, C. K., Kuang, X., Duan, F., Zhang, Z., Qi, H. J., Wang, T., Digital light processing 3D printing of conductive complex structures. *Additive Manufacturing*, 2017, **18**: pp. 74–83.

7

Additive Manufacturing Processes Utilizing Powder Bed Fusion Technique

7.1 Introduction

AM processes based upon powder bed fusion techniques (PBFs) were commercial pioneers and are mostly based upon the selective laser sintering (SLS) technique, which was originally employed for creating plastic prototypes via focussed or point-based laser scanning methods. Unlike vat photopolymerization processes, PBFs do not require any additional support material/structures.

This is the second chapter of Section B of this book, which presents process specific details of various AM processes. In the previous chapter, readers were introduced to all the necessary discussions on the vat photo polymerization process. This chapter provides details of AM techniques utilizing powder bed fusion processes including materials; powder fusion mechanism; process parameters and modelling; powder handling; powder fusion techniques, including solid state sintering, chemical sintering, complete melting, liquid phase sintering/partial melting, indirect processing, pattern method and direct sintering; powder bed fusion process variants including low temperature laser-based processing, metal and ceramic laser-based systems, electron beam melting (EBM) and line- and layer-wise systems; and strengths and weaknesses of PBF-based AM techniques. This chapter concludes the discussion with a summary.

7.2 Materials

Ideally PBFs can fabricate components from any material that can undergo melting and re-solidification. However, a special class of materials are best compatible with these PBFs. These mainly include:

- Thermoplastic materials, owing to low melting point, thermal conductivity, balling tendency

- Polyamide (PA) for plastic components of functional parts
- Glass-filled PAs to make parts extra strong and rigid with reduced ductility
- Polystyrene-based materials with lesser residual ash content in investment casting
- Amorphous polymer materials which tend to create porous structures
- Crystalline materials for high density, better surface as well as mechanical traits, but their shrinkage, curling and distortion exceeds that found in their amorphous counterparts
- Elastomeric thermoplastic polymers for high flexibility
- Biocompatible materials for specific uses
- Numerous proprietary metals, for example, RapidSteel, RapidSteel 2.0 by DTM Corp., LaserForm ST-100
- Metal alloys such as Ti–6Al–4V, steel alloys, CoCrMo, Inconel, etc.

Initially SLS was employed for plastic prototypes. However, today these processes are utilized for a broad range of metallic, polymeric, composite and ceramic raw materials.

PBF processes use a distinct method to fabricate metallic and ceramic parts. In case of metal parts, there are four main approaches, which include: (1) liquid phase sintering; (2) full melting; (3) indirect processing and (4) pattern methods. In case of ceramic components, there are four main processing approaches, including: (1) direct sintering; (2) chemical induced sintering; (3) indirect processing and (4) pattern methods.

7.3 Powder Fusion Mechanism

A few traits are common to each PBF process. These include:

1. A thermal source which enables powdered particles to fuse
2. A control mechanism to control fusion to a specified area of every layer
3. Mode of addition of subsequent layer
4. Mode to smoothen each layer before addition of next layer.

Understanding the mechanism of the SLS process, which is the basis of almost all PBF techniques, is an important part of comprehending the mechanism of PBF processes and is considered a baseline for comparison. Its working principle is therefore discussed in detail. A counter-moving roller set levels powder across the build area. This powder is of around 0.1 millimetre thickness. Fusion of thin layers of this powder takes place in the process of SLS. The part is built inside an insulated chamber with nitrogen atmosphere for preventing oxidation as well as powder material degradation. The temperature of the powder is slightly higher than its melting and/or glass transition temperature. Higher temperature around the component being fabricated is ensured with the help of an infrared or resistive heating device kept above the build platform. A similar infrared heating device is kept above the feed cartridge for pre-heating the powder before spreading it upon build area. Maintaining elevated temperatures at these two locations (above build platform as well as preheating point) amounts to considerable reduction in requisite laser power for powder fusion, as well as in part warping/curling owing to non-uniformity in the expansion and contraction. Upon completion of a desirable layer formation and required preheating, a concentrated carbon dioxide laser is directed towards the powder bed and its movement enables thermal fusion of the powdered material layer into the desired slice cross-section. The powder in the vicinity is not affected owing to the highly controlled laser beam. Surrounding powder, in fact, supports the proper build of the component.

Once a complete layer is formed, the build table lowers by an amount equal to that of a layer's thickness. Repetition of the entire process takes place till the fabrication of the complete component is accomplished. To enable removal of final component, some cool-off period is normally required. This is needed because of two reasons: (1) premature exposure of the part to ambient temperature will lead to warping because of unequal contraction; (2) to allow parts to uniformly arrive at a manageable handling temperature.

Finally, the components are taken out. After removal of loose remains of powder, the necessary finishing operations are performed before finalizing the parts. A schematic of the PBF process along with all the necessary subsystems is presented in Figure 7.1.

Various PBFs are obtained by changing the mode of fusion of powdered material. Four variable ways to obtain fusion can be: (1) sintering in solid state; (2) sintering in liquid state; (3) chemical binding; and (4) complete melting.

However, liquid phase sintering as well as melting are most prominently used. A brief description of the underlying principles of each of these fusion techniques is discussed over the subsequent sections.

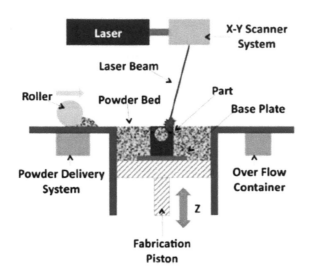

FIGURE 7.1
Powder Bed Fusion Schematic. (*Source:* [1]. Reprinted from Thompson et al., An overview of direct laser deposition for additive manufacturing. Part I: transport phenomena, modeling and diagnostics. *Additive Manufacturing*, 2015, 8: pp. 36–62, Copyright 2015, with permission from Elsevier.)

7.4 Process Parameters and Modelling

There are four main categories of process parameters with reference to PBF processes which are categorized as: (1) laser; (2) scan related; (3) powder; and (4) temperature related parameters. A detailed flowchart representing all these four is shown in Figure 7.2. A few important parameters are defined in Table 7.1. Apart from these, other relevant parameters are discussed below.

TABLE 7.1

Laser Sintering Process Parameters

Parameter	Description
Laser power	Applied power of laser as it scans area of each layer
Scan speed	Velocity at which laser beam travels as it traverses a scan vector
Scan spacing	Distance between parallel laser scans
Scan count	Number of times laser beam traverses a scan vector per layer
Scan strategy	Pattern of laser as it scans over a layer, in combination with the laser parameters used in each specific area
Layer thickness	Distance build platform lowers for spreading of new layer of powder
Build temperature	Temperature of process chamber and/or part bed

Source: [3]. Goodridge and Ziegelmeier, Powder bed fusion of polymers, in *Laser additive manufacturing*, M. Brandt, Editor, 2017, Woodhead Publishing, pp. 181–204.

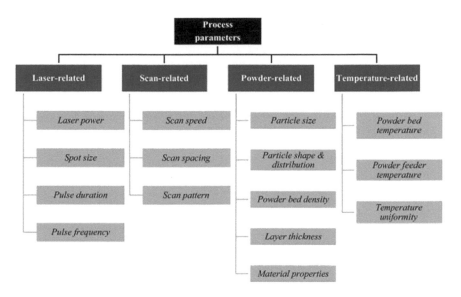

FIGURE 7.2
Principal Processing Parameters Involved in Selective Laser Melting Process. (*Source:* [2].
Aboulkhair et al., Reducing porosity in AlSi10Mg parts processed by selective laser melting.
Additive Manufacturing, 2014, 1–4: pp. 77–86, under Creative Commons Licence.)

7.4.1 Processing Temperatures ($T_{Processing}$)

$T_{Processing}$ is an important parameter in case of sintering in solid state. Normally in case of sintering, $T_{Processing}$ is kept in the range of half of $(M.T.)_{absolute}$ and M.T., where M.T. stands for melting temperature.

The diffusion rates are directly proportional to the temperatures. This is the reason behind the rapid increment in sintering rates as the temperatures approach the M.T. for a given material. This is modelled with a modified Arrhenius relationship.

7.4.2 Total Free Energy (E_s)

The product of total surface area of particle (S_a) and surface energy per unit surface area (γ_s) is called total free energy (E_s), which is another important parameter in the case of solid state sintering processes. This is, however, subject to constancy of material, temperatures and atmosphere. It can thus be mathematically expressed as $E_s = \gamma_s \times S_a$ for a given material, temperature and atmosphere.

This needs to be minimized for effective sintering. Fusion of particles at higher temperatures tends to lower surface area and, in turn, surface energy. This leads to reduced sintering rates and longer times, which are desirable to lower porosities to a remarkable degree.

7.4.3 Surface Area Bed/Volume Ratio of Particle ($SA_{bed}/VR_{particle}$)

The powder bed surface area to particle volume ratio is directly proportional to the force of free energy driving. The practical implication of this fact is that particles of small size will be subjected to increased driving force for necking as well as consolidation. This amounts to enhanced sintering at reduced temperatures for small particles as compared to their bigger counterparts.

7.4.4 Particle Dimensions

"Particle dimension" refers to the size of the particles used as raw material. Finer particles result in a smooth surface finish as well as accuracy in parts. However, they are extremely difficult to manage. Large particles on the other hand are easy to manage but restrain the feature size, accuracy, surface finish as well as least layer thickness.

7.4.5 Raw Material Thermal Properties

Most of the raw materials compatible with PBFs exhibit about 3–4% shrinkage, thereby increasing the tendency of the part to distort. Lower thermal conductivity is desirable in the raw materials. This is chiefly owing to the fact that it is possible to control the melt pool as well as solidification rates in case of lower thermal conductivity.

7.5 Powder Handling

There are a variety of techniques to handle powders in different PBF processes. The basic requirements of any effective powder delivery system are as below:

1. Capacity of powder reservoir should be enough to support entire build height without need to refill during the process.
2. Measured and exact amount of powder should be transferred from reservoir to build substrate.
3. Thin and even layer should be smoothly deposited over previous layers.
4. Spreading of current layer should not disturb previous layers due to excess force.
5. It should possess capability to render flowability (that is reduced owing to reduced particle dimensions) to powder for effectiveness in delivery.

6. It should be able to provide inert atmosphere to powders to keep a check upon powder reactivity and consequent damages.
7. It should limit formation of airborne particles that can seriously hamper the effectiveness of related AM systems.
8. It should be able to provide as small particles as feasible while effectively dealing all the above issues for good quality parts.

There can be different systems of powder delivery. The first one most commonly employs dual-feed cartridges and is used in SLS systems. Another utilizes a doctor blade. Yet another powder delivery system uses a hopper feeding mechanism. Ultrasonic vibrations enhance the efficiency of both the doctor blade and hopper feeding system. For multi-material processing, a discrete hopper is used for each material.

The recycling of the adjoining powder is also an important issue that needs to be taken care of. Due to constant reheating of particles adjacent to the part, they tend to fuse, as well as vary in their chemical properties. Their molecular weight also undergoes changes. The recycling can be either frequent or restricted depending upon the changes in powder materials. If there is negligible change in powder properties, then they can be frequently recycled. However, if there are drastic changes in powder, then control methods need to be adopted. One such method is mixing of used and unused powders. For example, mixing equal amounts of unused, overflow/feed and build platform powders. This method can sometimes lead to inconsistencies. Another approach to recycling issues can be on the basis of the MFI (melt flow index) of powder. This is a better method than mixing in fractions. Normally a fraction of the used powder is always wasted. The cost of the process increases with increased requirement of recyclability of powders.

7.6 Powder Fusion Techniques

7.6.1 Solid State Sintering

Fusion of thermally processed material powders via various routes is one of the earliest trends in AM. An important fact to be understood is that even at temperatures near to M.T., sintering based upon diffusion is the slowest fusion process amongst the PBF techniques. Since process speed is an important parameter of estimating effectiveness of an AM process and particle fusion time for sintering is always higher than that for melting, very few techniques use sintering as primary fusion process. During solid state sintering, no melting should take place. This in turn implies that this is a solid state processing technique. Figure 7.3 shows the various phases undergone by powder particles during solid state sintering: starting from

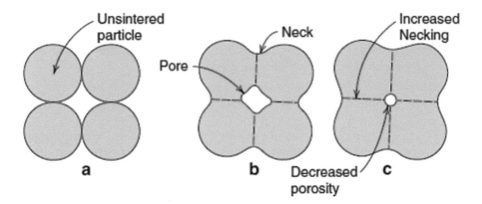

FIGURE 7.3
Solid State Sintering: (a) Closed Packing of Powder Particles before Sintering; (b) Agglomeration of Particles at Temperatures above Half Absolute Melting Points; (c) Increased Neck Size and Reduced Porosity Pore Sizes. (*Source:* [4]. Reprinted/adapted from Gibson et al., Powder bed fusion processes, in *Additive manufacturing technologies: 3D printing, rapid prototyping, and direct digital manufacturing*, 2015, Springer London, with permission from Springer Nature.

their close packing prior to sintering, agglomeration at temperatures greater than $(M.T.)_{1/2}$ and final fusion with increase in neck size and reduction in pore sizes.

Despite the fact that fusion times in sintering are higher, it is an important thermal process since it affects part build in more than one way. These ways include:

1. Fusion of particles due to sintering

2. During sintering of loose powder at increased temperatures of powder build platform, its tensile and compressive strength will be gained and thus curling is minimized. This, however, has a limitation also since particles are agglomerated and thus their grain size increases.

3. Unlike the cases where rapid fusion occurs, sintering leads to porosity-free and fully dense metal parts.

4. Part growth and thick part skin formation takes place due to solid state sintering which increases with increase in part dimensions and thus sintering time. This is mainly because of the heating up of loose powder surrounding the part.

7.6.2 Chemical Sintering

Chemical or chemically induced sintering is mainly used for ceramics. Here, chemical reaction between powders/powder+atmospheric gases leads to formation of products that bind powders. For example, if silicon carbide is processed using laser in the presence of O_2, silicon dioxide is formed that binds silicon dioxide and its carbide. Similarly, ZrB_2 in the presence of O_2 and aluminium in the presence of N_2 can be processed. It has been established that combinations of high temperature ceramics and/or intermetallic powders can be made to react using lasers. Since a lot of energy is released during the process, so the requirement of laser power is considerably reduced. This process has limited usage owing to high post-processing requirements and corresponding cost for treatment of porosities.

7.6.3 Complete Melting

In the complete melting process, melting of the material region more than layer thickness takes place due to heat energy. Laser/electron beam energy used in such processes has enough power to melt part of the previously solidified layer apart from the presently scanned one. This process therefore results in full density, bonded parts of engineering metals and their alloys, as also semi-crystalline polymeric materials. Nylon, titanium, stainless steels, etc. are widely used as PBF raw materials. However, full melting also amounts to part growth which is generally an undesirable output. Balanced parameters therefore need to be chosen carefully for optimal strength and accuracy.

7.6.4 Liquid Phase Sintering/Partial Melting

In liquid phase sintering, one of the powders is in a partially molten state thereby acting as a gluing material for another one which is a solid. It is the most widely utilized binding technique owing to its high-quality results. This technique also finds wide utilization in powder metallurgy applications. The partial melting technique is used in more than one way for AM fusion applications. A classification was proposed by Kruth et al. [5] to distinguish between various techniques:

- Separate binder and structural materials that can be composite or coated particles
- Indistinct binder and structural materials.

7.6.5 Indirect Processing

Indirect processing uses either metal powders with coating of polymers or a mix of metal and polymer particles. The polymer particles melt and bind the solid metal powder particles. After this, the polymer glued green parts are treated in a furnace which occurs in two stages. The first stage is debinding, in which vaporization of polymer particles takes place and metallic particles also experience partial sintering. In the second stage, the part is either infiltrated using a low melting point metal or consolidated to reduce porosity, since the parts fabricated via this technique are generally porous. This is followed by further processing to enhance density of parts. This indirect processing technique is generally used for metal as well as ceramic parts. In metals, it results in composite fabrication when infiltrated with some metal and single metal parts when consolidated. In ceramics, if the metal powders are used for infiltration then ceramic/metal composites are obtained; if molten silicon is used then ceramic matrix composites are created owing to reaction between silicon and remaining carbon to form SiC.

7.6.6 Pattern Method

This approach is also used for metals as well as ceramic parts fabrication. Unlike the use of metallic powders for part creation, this technique uses a pattern for creation of parts. Sand casting molds (mixture of sand and thermosetting binder) and investment casting patterns (polystyrene/wax-based powders) mainly use patterns made via this route.

7.6.7 Direct Sintering

This includes maintaining high powder bed temperatures and utilization of laser for accelerating its sintering at specified layer position. The resulting parts have high porosities and thus require post-processing in furnace for densification.

All the PBF processes use any combination of techniques described above for binding purposes. The choice of technique is based upon raw materials, accuracy, strength and amount of energy needed to fabricate the PBF part.

7.7 Powder Bed Fusion Process Variants

Based upon differences in powder delivery techniques, heating methods, ambient conditions, energy inputs, optics and remaining important features, many variants of PBF processes have evolved. These mainly include:

1. Low temperature laser-based processing
2. Metal and ceramic laser-based systems
3. Electron beam melting (EBM)
4. Line- and layer-wise systems.

7.7.1 Low Temperature Laser-Based Processing

Low temperature laser-based systems process polymers directly and metals/ceramics indirectly. Low temperature systems that process polymers are known by two names: laser or selective laser sintering, abbreviated as LS and SLS respectively. SLS Sinterstation 2000, a pioneer PBF modeller, was launched by the DTM Corporation. DTM Corp. was later taken over by 3D Systems. Most of the advanced machines based on this processing still utilize SLS principles with newer improved features. Since these machines are based on carbon dioxide lasers and inert nitrogen gas atmosphere, they cannot directly process pure metals/ceramics. They, however, process polymers, specially nylon polyamides, polystyrene-based elastomeric material, etc. optimally. Metals with a polymeric binder are also indirectly processed by them. Sinterstation Pro systems are a recent addition to this category of modellers. Another leading technology provider of this technique is EOS GmbH. Their machines are designed in a manner to use only one material type per machine. EOSINT P (plastic parts, 1991) was their first machine and the next one was EOSINT M250 (foundry sand mold, 1995). EOSINT M250 Xtended (metal powders, specially mixture of bronze and nickel, 1998) was the next addition utilized in direct metal laser sintering (DMLS) based on liquid state processing. Subsequently, several other variants of DMLS came with advanced capabilities of raw material choice, multiple platforms, multiple lasers, etc. A schematic of SLS is shown in Figure 7.4 and of a DMLS system is given in Figure 7.5.

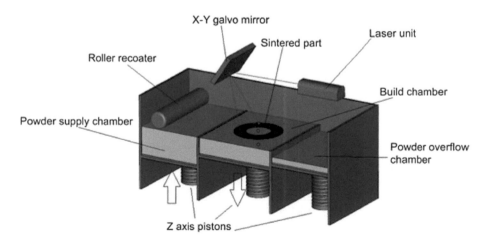

FIGURE 7.4
Schematic of SLS. (*Source:* [3]. Reproduced with permission from Goodridge and Ziegelmeier, Powder bed fusion of polymers, in *Laser additive manufacturing*, M. Brandt, Editor, 2017, Woodhead Publishing, pp. 181–204.)

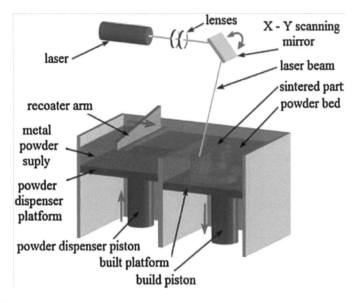

FIGURE 7.5
Schematic of DMLS. (*Source:* [6]. Reprinted from Singh et al., Material issues in additive manu-
facturing: a review. *Journal of Manufacturing Processes*, 2017, 25 (Supplement C): pp. 185–200,
Copyright 2017, with permission from Elsevier.)

7.7.2 Metal and Ceramic Laser-Based Systems

Germany-based EOS GmbH, UK-based MTT Technologies, Germany-
based Concept Laser GmbH and France-based Phenix Systems are four
main technology providers in this genre. Though selective laser melting
(SLM) is the main processing technique under this category, terminology
like LaserCUSING and DMLS is also utilized. The schematic for the SLM
working principle is given in Figure 7.6. Early research in the field of
SLM was mostly unsuccessful. Present versions of the SLM system are
basically variants of SL powder remelting (SLMR) (FILT, Germany). Here
Nd-YAG instead of carbon dioxide lasers were used since they possessed
better tuning to metal absorptivity. Now SLM systems mostly use fiber
lasers which are even better and more economical as compared to
Nd-YAG lasers. Scan patterns, f-theta lenses, and inert atmospheres with
low oxygen characterize SLM machines today. Another important feature
is rigidly attaching the part to a base plate at the build platform bottom
to prevent distortion, thereby slightly restricting the design flexibility.
With the introduction of M270 DMLS (2004) which can handle titanium
and steel alloys, CoCrMo, etc., EOS GmbH became market leaders in
SLM technology.

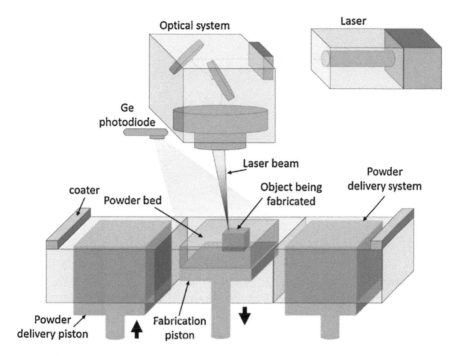

FIGURE 7.6
Schematics of Powder Bed Fusion Equipment: Selective Laser Melting. (*Source:* [7]. Bisht et al., Correlation of selective laser melting-melt pool events with the tensile properties of Ti-6Al-4V ELI processed by laser powder bed fusion. *Additive Manufacturing*, 2018, 22: pp. 302–306, under Creative Commons Licence.)

Concept Laser GmbH markets machines as LaserCUSING (cladding + fusing) systems. This is widely used for injection molding and tooling applications and alloys of stainless, work-hardened steels, aluminium alloys, etc. Phenix Systems hold expertise in building SLM platforms that can benefit dental industries (small, intricate parts) to extremely versatile systems (metals and ceramics). These systems are specially suitable to process high melting point materials. 3D Micromac AG (Germany) develops small-scale SLM systems. They use fine powdered materials, fiber lasers and a twin material feeding mechanism, and are used to produce fine, intricate parts.

7.7.3 Electron Beam Melting

Electron beam melting (EBM) has gained enormous success as a prominent PBF process and utilizes high energy electron beams for inducing fusion of metal particles. Arcam AB (Sweden, 2001) made it commercial. EBM works on the principle of scanning across a thin layer of already spread powder by a focussed electron beam. A real-time image of an EBM machine and its schematic is presented in Figure 7.7. This is very similar to SLM where

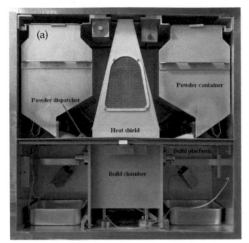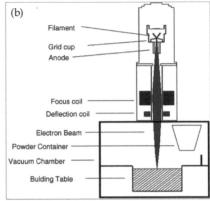

FIGURE 7.7
Real Time Image of EBM Machine and its Schematic: (a) Fusing on an EBM Machine; (b) Scheme of an Additive Machine. (*Source:* [8]. Reproduced with permission from Petrovic et al., Additive layered manufacturing: sectors of industrial application shown through case studies. *International Journal of Production Research*, 2011, 49(4): p. 1061–1079.)

scanning is done with lasers. However, there are many points of difference between SLM and EBM. First, EBM uses electrons as an energy source while SLM uses a laser. Second, the materials processed using EBM are conducting metals, whereas they can be polymers, metals or ceramics in SLM. Third, EBM needs a vacuum while SLM needs an inert atmosphere. Fourth, scanning is done with deflection coils in EBM and galvanometer in SLM. Fifth, EBM uses an electron beam to preheat powders while SLM uses infrared heaters. Sixth, with EBM the surface finish is moderate to poor but it is moderate to excellent in the case of SLM components, and so on. One bright prospect of EBM systems lies in their instantaneous beam movement capability. With improvement in scanning strategies, EBM efficiencies will increase significantly.

7.7.4 Line- and Layer-Wise Systems

PBF approaches have great flexibility but high cost owing to localized melting of material accomplished either by electron beams or lasers, making these systems quite expensive. Development of newer processes that process a layer or a line has been an area of interest. Techniques for line- and layer-processing of polymers emerged less than a decade ago. Three such techniques based upon use of infrared fusion energy are: (1) mask-based sintering; (2) printing an absorptivity increasing medium upon part area; and (3) printing a sintering inhibiting agent on outer side of part area.

Examples of these systems include selective mask sintering (Sintermask GmbH, Germany, 2009), ZORRO (Sintermask GmbH, Germany), high-speed sintering (Loughborough University, UK), selective inhibition sintering, inkjetprinting plus sintering (fcubic AB, Sweden), etc.

7.8 Strengths and Weakness of PBF-based AM Techniques

There are many strengths of PBF-based AM techniques. These mainly include:

- Ability of powder bed to act as support material in polymer-based PBFs leading to appreciable time saving in fabricating and post-processing parts.
- Ability to incorporate complex interior features like cooling channels, etc.
- Ability to fabricate highly intricate parts, especially with the aid of smaller size particles.
- In most modern PBF machines, preheating and cooling cycles are undertaken separately, thereby controlling the build times with the aid of removable build platforms.
- PBF machine productivity is appreciably high and can be compared to other AM processes.
- Nesting of parts in case of polymer-based PBF processes is very easy.
- Cost, productivity and build up times in case of polymer-based PBF techniques are appreciably reduced owing to the ability to nest parts without need of any support structures.

PBFs are, however, not free from their own set of limitations. A few of these include:

- Requirement of support structures in case of metal PBFs for limiting the component warping during processing.
- Porosity is an issue with most PBFs thereby necessitating post-processing in the form of infiltration/high temperature sintering to obtain full densities.
- Post-processing of metallic parts is costly and consumes a lot of time.
- Solid-based PBF process leads to less accurate parts possessing inferior surface finish as compared to liquid-based techniques.
- Spreading and handling fine particles is an arduous task.

- In PBFs, lower thermal conductivities of materials is desirable but this leads to reduction in part growth rate and in turn increased build time. This will in turn enhance the cost of PBFs.
- Total build up times are higher than most other AM techniques owing to involvement of preheating and cooling cycle times.
- Most modern PBF machines are expensive.

7.9 Summary

PBF techniques are amongst the oldest AM processes. LS is an extensively popular technique for polymer component fabrication. Like other AM techniques, PBFs are also most suitable for low to medium batch production. Metal-based PBFs based upon laser and electron beams are in huge demand today. They have numerous biomedical, aerospace and automotive applications. This is mainly owing to their ability to fabricate parts with intricate geometries and enhanced material desirable characteristics like strength as well as aesthetics. Some major current areas of research involving PBF technology are presented in Figure 7.8.

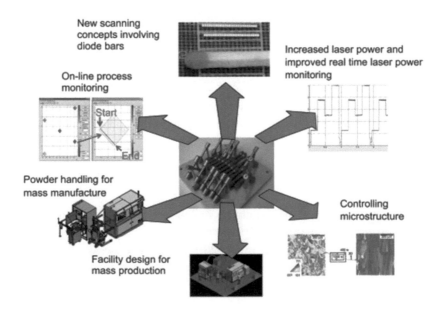

FIGURE 7.8
Current Major Areas of Research Involving Powder Bed Technology. (*Source:* [9]. Brandt, The role of lasers in additive manufacturing, in *Laser additive manufacturing*, M. Brandt, Editor, 2017, Woodhead Publishing, pp. 1–18, under Creative Commons Licence.)

With ever increasing growth and newer innovations in the field of AM, these processes are bound to become more versatile. Also, the accompanying cost in terms of time, price, quality would decrease which would increase their competitiveness.

In this chapter, an attempt has been made to cover all the necessary aspects of the powder bed fusion AM technique. The next chapter covers extrusion-based AM processes.

References

1. Thompson, S. M., Bian, L., Shamsaei, N., Yadollahi, A., An overview of direct laser deposition for additive manufacturing. Part I: transport phenomena, modeling and diagnostics. *Additive Manufacturing*, 2015, **8**: pp. 36–62.
2. Aboulkhair, N. T., Everitt, N. M., Ashcroft, I., Tuck, C., Reducing porosity in AlSi10Mg parts processed by selective laser melting. *Additive Manufacturing*, 2014, **1–4**: pp. 77–86.
3. Goodridge, R., Ziegelmeier, S., Powder bed fusion of polymers, in *Laser additive manufacturing*, M. Brandt, Editor, 2017, Woodhead Publishing, pp. 181–204.
4. Gibson, I., Rosen, D. W., Stucker, B., Powder bed fusion processes, in *Additive manufacturing technologies: 3D printing, rapid prototyping, and direct digital manufacturing*, 2015, Springer, pp. 107–145.
5. Kruth, J. P., Mercelis, P., Van Vaerenbergh, J., Froyen, L., Rombouts, M., Binding mechanisms in selective laser sintering and selective laser melting. *Rapid Prototyping Journal*, 2005, **11**(1): pp. 26–36.
6. Singh, S., Ramakrishna, S., Singh, R., Material Issues in additive manufacturing: A review. *Journal of Manufacturing Processes*, 2017, **25**(Supplement C): pp. 185–200.
7. Bisht, M., Ray, N., Verbist, F., Coeck, S., Correlation of selective laser melting-melt pool events with the tensile properties of Ti-6Al-4V ELI processed by laser powder bed fusion. *Additive Manufacturing*, 2018, **22**: pp. 302–306.
8. Petrovic, V., Vicente, H. G. J., Jordá, F. O., Delgado, G. J., Ramón, B. P. J., Portolés, G. L., Additive layered manufacturing: sectors of industrial application shown through case studies. *International Journal of Production Research*, 2011, **49**(4): p. 1061–1079.
9. Brandt, M., The role of lasers in additive manufacturing, in *Laser additive manufacturing*, M. Brandt, Editor, 2017, Woodhead Publishing, pp. 1–18.

8

Additive Manufacturing Processes Utilizing an Extrusion-Based System

8.1 Introduction

Extrusion-based AM processes and their details are covered in this chapter. These utilize the basic principle of forcing pressurized semi-molten material out from a nozzle at either continuous (constant layer thickness) or variable rates (variable layer thickness) to obtain layers after their complete solidification. Bonding of these layers occurs till the complete artefact is obtained. There are two basic control mechanisms to enable layer formation. One of these is temperature-based and another is chemical change-based. In temperature-based systems, liquification of molten material occurs in the reservoir to enable its flow through the nozzle and subsequent bonding with the previously deposited layer/substrate. In chemical-based systems, solidification of layers is driven by a chemical change due to curing of residual solvents or reaction with air or simply drying. This process has mainly biochemical applications and a very few industrial ones.

This is the third chapter of Section B of this book, which presents process specific details of various AM processes. In the previous chapter, readers were introduced to all the basic details of the powder bed fusion AM technique. This chapter gives details of AM processes utilizing the extrusion-based system including: basic principles of extrusion-based processes; fused deposition modelling including its performance measures and FDM limitations; bio-extrusion; contour crafting; non-planar systems; RepRap FDM systems; and applications. This chapter concludes the discussion with a summary.

8.2 Basic Principles of Extrusion-Based Processes

A variety of basic steps characterize any extrusion-based process, including material loading which is done from a chamber and liquefying by means of heat, pressurizing, extruding from a nozzle, plotting, its bonding and finally incorporating support structures. Newtonian fluid theory is applied for modelling most extrusion processes.

One way of classifying extrusion-based AM techniques is material melting, based upon which there are two subgroups. Material melting takes place in the first kind of process, while there is no material melting in the second subgroup. The detailed classification on this basis along with the relevant processes is presented in Figure 8.1.

Different types of nozzles can be utilized in these systems, including those actuated by pressure, volume driven injection, solenoid, piezoelectric, filament driven wheels, mini-screw extruder, etc.

Another classification can be based upon the type of extruder used, i.e. plunger-, filament- and screw-based extrusion systems. Plunger-based extruders utilize a special metal/ceramic rod with thermoplastic binders. Filament-based extruders form the basis of fused filament machines (FFFs) and are based upon the use of filament. Screw-based systems use a screw which has a solid conveying zone and a metering zone. They are used where the other two types of extruders are not easy to apply. Figure 8.2 shows the three different types of extruders used in extrusion-based systems.

Fused deposition modelling is the most common extrusion-based process and is described in detail in the next section.

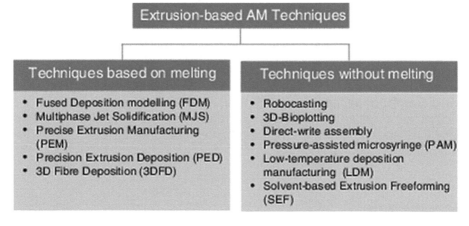

FIGURE 8.1
Classification of Extrusion-Based Systems on Basis of Material Melting. (*Source:* [1]. Reproduced with permission from Vaezi et al., Multiple material additive manufacturing. Part 1: a review. *Virtual and Physical Prototyping*, 2013. 8(1): pp. 19–50.)

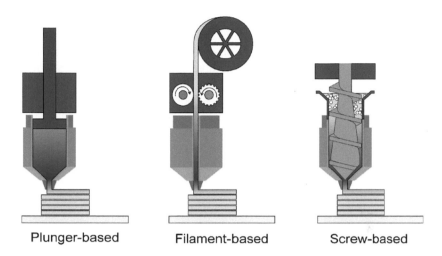

| Plunger-based | Filament-based | Screw-based |

FIGURE 8.2
Different Types and Approaches for Extrusion-Based Additive Manufacturing. (*Source:* [2]. Gonzalez-Gutierrez et al., Additive manufacturing of metallic and ceramic components by the material extrusion of highly-filled polymers: a review and future perspectives. *Materials*, 2018, 11(5): p. 840, under Creative Commons Licence.)

8.3 Fused Deposition Modelling

FDM, which was initially proposed by S. Scott Crump around 1980s and commercialized by Stratasys, is an extremely popular and robust AM technique since complicated parts can be obtained in reasonable time. Like other AM techniques, it requires no tooling and minimal human interference. This can be used to obtain models, prototypes as well as end parts and is based on the layered AM principle. Its working involves uncoiling of plastic filament from its spool which then goes into an extrusion nozzle. The choice of nozzle and filament depends upon specific need. Melting of material occurs due to the heating element in the nozzle. Horizontal and vertical direction movement occurs with the help of an automated computerized mechanism, which is under the direct control of a CAM software package. Layer formation takes place by extruding specific thermoplastic material directly either on substrate or on the previous layer.

8.3.1 FDM Materials

The most common FDM raw material is ABS (acrylonitrile butadiene styrene) plastic which is a carbon chain copolymer and is obtained by dissolution of butadiene styrene copolymer into a mixture of acrylonitrile and styrene monomers. Acrylonitrile provides heat resistance; butadiene provides impact strength, while styrene makes parts rigid. The principle of diffusion welding is utilized which is non-continuous owing to which material is not uniformly distributed. This reduces part strength and additionally amounts to anisotropic FDM parts.

Polycarbonates, polyphenyl sulphones, polycaprolactone, waxes, etc. are also used. Water soluble support materials are utilized with FDM (Stratasys Waterworks) which quickly dissolves by means of metal agitators with sodium hydroxide solvent.

8.3.2 Working Principles of FDM

A schematic of FDM is shown in Figure 8.3; it is based on the following working steps:

1. Exporting 3D solid model into FDM Insight™ in .stl format

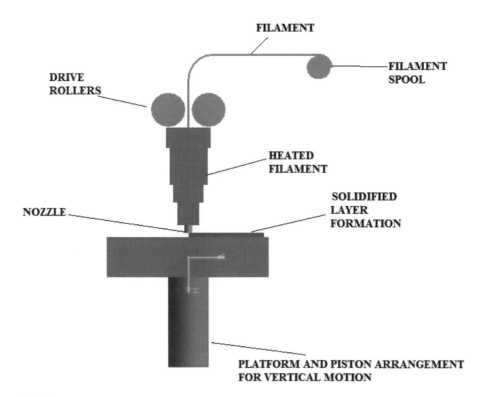

FIGURE 8.3
Schematic of FDM process. (Source. [3]. Reproduced from Srivastava, *Some studies on layout of generative manufacturing processes for functional components*, 2015, Delhi University, India.)

2. Generation of process plan for controlling FDM modeller
3. Feeding ABS filament into heating element to bring it into semi-molten form
4. Feeding filament via nozzle and its deposition upon previous layer/ substrate
5. Repeating process to obtain final part.

Since extrusion is in a semi-molten state, fusion of newer material into previous layers takes place. Movement of head around the x–y plane is followed by corresponding material deposition as per part geometry which in turn is followed by lowering of platform to enable deposition of new layer. This process continues till the complete part as per CAD data is obtained.

8.3.3 FDM Process Parameters

There are many FDM specific parameters which are presented in Figure 8.4.

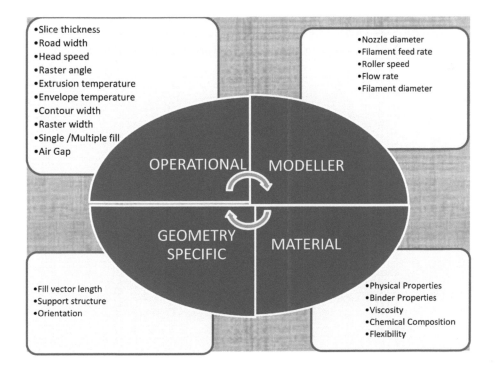

FIGURE 8.4
Various FDM Process Parameters. (*Source:* [3]. Reproduced from Srivastava, *Some studies on layout of generative manufacturing processes for functional components*, 2015, Delhi University, India.)

Some important process parameters are briefly introduced with their Insight corollary definitions below. These are also shown in Figure 8.5 for clarity.

- *Raster width:* "Raster width is the material bead width used for rasters. Larger values of raster width will build a part with a stronger interior. Smaller values will require less time and material."
- *Contour width:* "Contour width is the material bead width used for contours. Smaller values of contour width will build a part with a better surface finish. Larger values will require less time and material."
- *Air gap:* "Air Gap sets the distance between the part and the supports when creating containment supports. Large values might position the support curve too far from the part to give support, while values that are too small might embed the support curves in the part."
- *Raster angle:* "This angle is for rasters on the bottom part of layer. The angle is measured from the X axis of the current layer to the raster. Increase in raster angle would lead to more material consumption."
- *Slice height:* "The slice operation computes part curves by analysing cross-sections of the .stl file. It begins at the bottom of the model and progresses sequentially to the top at a constant interval known as slice height. Its value is based on the material and tip size used in modeler."

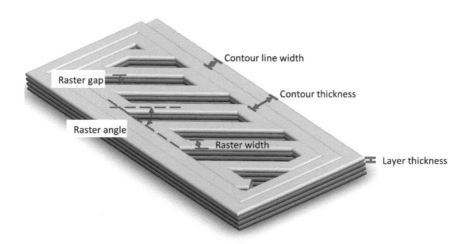

FIGURE 8.5

Illustration of Process Parameters in Extrusion-Based Processes. (*Source:* [4]. Reproduced with permission from Goh et al., Process–structure–properties in polymer additive manufacturing via material extrusion: a review. *Critical Reviews in Solid State and Materials Sciences*, 2019: pp. 1–21.)

- **Orientation:** "Part build orientation or orientation refers to the inclination of part in a build platform with respect to X, Y and Z axis. Where X and Y axes are considered parallel to build platform and Z axis is along the direction of part build."

8.3.4 Performance Measures

There can be many performance measures as briefly introduced below:

- **Build time:** Build time (BT) is the duration spent on an AM machine/modeller in absence of any bottlenecks. It should be clearly understood that BT is different from process time or speed despite one being used as a direct indicator of the other. This can be seen in Figure 8.6 which clearly shows BT as a subset of process time.

  In Figure 8.6 input collection or file preparation time constitutes preparing CAD model, STL generation, as well as time needed to interface. System preparation time includes time to orient model, generate support, select layer thickness, apply build styles, prepare machine, load or swap material and warm up machine. Post-build/post-processing is the sum of duration of post-build idle, part drainage/cool-off, chamber cool-off, cleaning, post-cure, removing supports, removing excess material, as well as final finish.

  A very important consideration is that superior parts might be desirable even at considerably high BT requirement. Lower BT may thus not be always preferable even when it means lower production cost in terms of money and time. Judicial weighing of BT against other design targets is thus mandatory before any final decision.

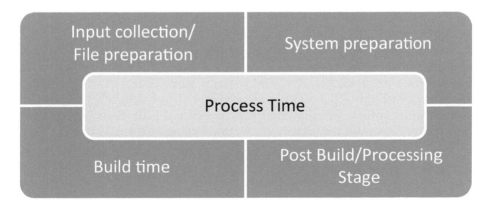

FIGURE 8.6
Components of Overall Process Time. (*Source:* [3]. Reproduced from Srivastava, *Some studies on layout of generative manufacturing processes for functional components*, 2015, Delhi University, India.)

- *Model material volume:* The Insight corollary defines this as: "Model material volume (MV) is the amount of raw material used for making the component." It is a matter of fact that most AM modellers, including FDM, utilize expensive MVs and their optimal utilization if of critical importance. Also, MV requirement is different for every application. For example, porous scaffolds need minimal MV but service model parts need more; concept models might need lesser MV but structural parts require fully dense parts; and so on.

- *Support material:* The Insight corollary defines this as "Support material (SM) volume is the amount of material used as a support for making a component." Significant amount of saving can be obtained by selecting optimal SM volume.

- *Production cost:* Production cost (PC) has the following components:

  1. Machine cost = (modeller cost + labor cost + running cost + annual maintenance cost) per hour × build time

  2. Support material cost = (cost of support material spool/volume of material per spool) × volume of support material used

  3. Model material cost = (cost of model material spool/volume of material per spool) × volume of model material used

  4. Total production cost = (1) + (2) + (3).

Apart from these, mechanical properties, including tensile, compressive and flexural strengths; surface roughness; dimensional accuracy; transverse strength; hardness; creep characteristics; etc. are also very important and can be similarly understood.

8.3.5 FDM Modellers

Two FDM modellers (Maxum Modeler powered by Insight v6.2 and Fortus 250mc powered by Insight v9.0) and their details are presented in this section. Apart from these two, numerous other FDM modellers with various traits are available. The FDM Maxum Modeler and its work volume are shown in Figures 8.7 and 8.8 respectively. The Fortus 250mc Modeler and its work volume are shown in Figures 8.9 and 8.10 respectively. Also, Figure 8.10 illustrates the primitives obtained using a Fortus 250mc.

8.3.6 FDM Limitations

Stratasys' FDM modellers have gained huge popularity owing to their simple operation and robust parts. However, they suffer from a number of inherent limitations, including restrictions on build speeds, accuracy and density. Though appreciably less layer thickness can be achieved, this is possible only with extremely high-end and expensive modellers. Sharp

FIGURE 8.7
FDM Maxum Modeler.

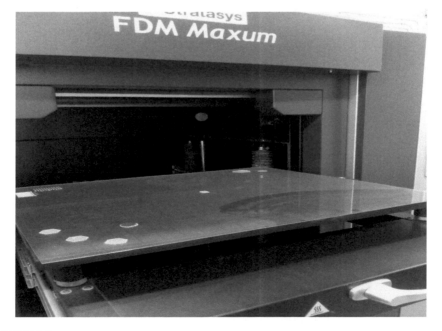

FIGURE 8.8
Build Chamber of FDM Maxum.

FIGURE 8.9
Fortus 250 mc.

FIGURE 8.10
Build Chamber of Fortus 250mc.

corners and interiors are very difficult to achieve owing to the roundness of the nozzle. FDM components exhibit high anisotropy and hence need to be carefully designed and used.

8.4 Bio-Extrusion

This is a technique for creating biocompatible and degradable scaffolds which have the unique characteristics of acting as a host to livings cells to enable tissue formation. These are characterized by presence of both micro (for growth of cells) as well as mini (for cell adhesion) pores. Hydrogel usage is common in scaffold creation since it provides a highly non-toxic biocompatible and conducive atmosphere to the growth of cells. Hydrogels are a class of water insoluble polymers that disperse in water and can be used to obtain rather weak scaffolds. To obtain stronger scaffolds, melt extrusion-based FDM techniques are utilized. However, these are inferior to gel-based ones owing to their lesser biocompatibility and biotoxicity. Figure 8.11 illustrates examples of scaffolds created by bio-extrusion.

8.5 Contour Crafting

This technique suggests use of a scraping tool to smooth surface midway, with the objective of contouring surfaces. This method is used to obtain considerable large parts in relatively lesser times. A real life example is shown in Figure 8.12 which shows a big machine as well as a scraping device used in contour crafting.

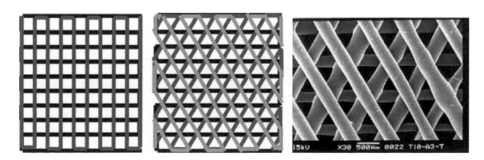

FIGURE 8.11
Images of Various Scaffold Designs Having Porous Structure along with Actual Image of Scaffold Developed Using Bio-Extrusion System. (*Source:* [5]. Reprinted from Zein et al., Fused deposition modeling of novel scaffold architectures for tissue engineering applications. *Biomaterials*, 2002, 23(4): pp. 1169–1185, Copyright 2002, with permission from Elsevier.)

FIGURE 8.12
Subsystems of Contour Crafting Developed at USC. (*Source:* [6]. Reprinted from Gibson et al., Extrusion based systems, in *Additive manufacturing technologies: rapid prototyping to direct digital manufacturing*, 2010, Springer U.S., with permission from Springer Nature.

8.6 Non-Planar Systems

Shape deposition, ballistic particle and curved laminated object-manufacture techniques come under the class of non-planar systems in which the layers are not plane or stratified. These techniques are used where additional surface toughness is required.

8.7 RepRap FDM Systems

Reprap is an open source technology based upon FDM machines. It has numerous design and material variants. A real life example is shown in Figure 8.13.

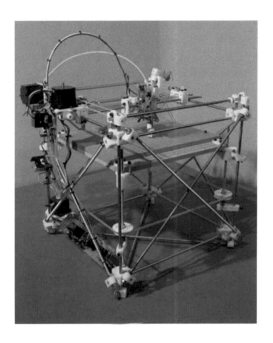

FIGURE 8.13
Example of FDM-Based RepRap Set-Up. (*Source:* [6]. Reprinted from Gibson et al., Extrusion based systems, in *Additive manufacturing technologies: rapid prototyping to direct digital manufacturing*, 2010, Springer U.S., with permission from Springer Nature.

8.8 Fab@home FDM Systems

This is a low-cost concept utilizing laser cut polymer sheet frames assembled together in a prespecified fashion.

8.9 Applications

Typical applications of FDM include concept models, fit and form models, models to be used for indirect AM manufacturing applications, investment casting, injection molding, etc. These are generated quickly at economical rates. There are no toxic chemicals involved in the FDM process.

FDM applications can also be utilized to obtain metal, ceramic, multi-material as well as metal-ceramic components with the help of highly-filled polymeric materials (HPs). This can be accomplished by the process of

HP-FDM and multi-material FDM techniques. However, parts need to be debound and sintered in this case. Multi-material parts can be effectively fabricated by this route. Bio-ceramic as well as composite scaffolds and three-dimensional lattice structures can be obtained by the solvent-based free-forming technique. Multi-material extrusion-based systems have been utilized to produce PLGA-collagen scaffolds. A hybrid system of FDM with Robocasting has proven abilities of enhanced multi-material part fabrication and can be used to print conductive circuits on components. A wide variety of biocompatible materials can be obtained using extrusion-based techniques. Devices like an RP robot dispensing for chitosan-HA scaffolds and a bio-extruder for tissue engineering (TE), as well as polycaprolactone (PCL) scaffold fabrication, have been introduced by various researchers. Scaffolds for soft tissue engineering and hydrogels can be fabricated with the help of extrusion-based 3D bio-plotting process. Hybrid bio plotting can fabricate solid biodegradable parts with hydrogels.

A few applications of extrusion-based AM processes are shown in Figures 8.14 and Figure 8.15. Figure 8.14 shows real-time primitive parts fabricated using the Fortus 250mc modeller and Figure 8.15 shows the parts ready for removal.

FIGURE 8.14
Components Made Using FDM Technique. (*Source:* [3]. Reproduced from Srivastava, *Some studies on layout of generative manufacturing processes for functional components*, 2015, Delhi University, India.)

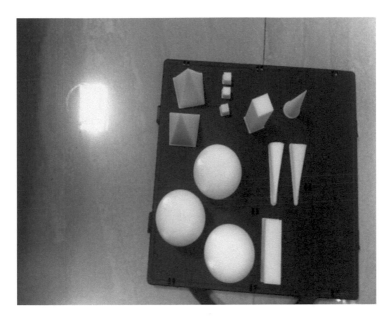

FIGURE 8.15
Primitives Ready for Removal. (*Source:* [3]. Reproduced from Srivastava, *Some studies on layout of generative manufacturing processes for functional components*, 2015, Delhi University, India.)

8.10 Summary

Extrusion-based processes are common techniques for fabricating thermo-plastic parts. Feedstock materials, binding mechanisms, monitoring systems, filaments, nozzle design, building strategies, etc. are a number of aspects that need careful planning and attention. Simulation tools need development in this field for reducing cost and enhancing effectiveness. Surface roughness needs to be minimized and mechanical characteristics need to be improved for parts made by these processes. Apart from hardware, development of good quality pre-processing tools is also important. Tool path planning is also important.

Figure 8.16 shows a self-explanatory integrated approach of extrusion-based systems which clearly shows material development, process control, microstructural characterization and design, modelling and simulation of data as the four main domains of control.

HP-extrusion is an extremely important and innovative aspect of extrusion-based AM techniques and can be utilized for fabrication of a wide variety of material parts. Development of simulation techniques related to debinding and sintering can go a long way in increasing efficiency of these systems.

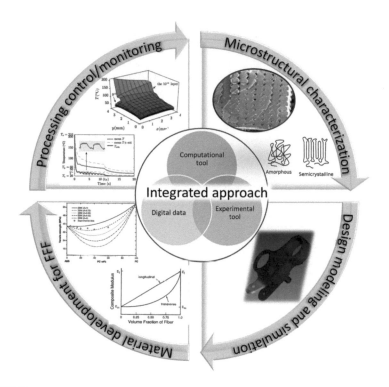

FIGURE 8.16
Integrated Approach of Various Aspects for Extrusion-Based AM Techniques. (*Source:* [4].
Reproduced with permission from Goh et al., Process–structure–properties in polymer additive manufacturing via material extrusion: a review. *Critical Reviews in Solid State and Materials Sciences*, 2019: pp. 1–21.)

There are many gaps and scope for future research in the field of extrusion-based processes. These include a need to establish a meaningful comparison of various techniques for a standardized part. This would involve systematic investigation of AM techniques. There are very few standardization techniques available, which leads to a huge shortcoming in the field of AM techniques. This can be another broad area of research. Another important area requiring extensive research is the diagnostic techniques aspect required at various stages of part fabrication of AM parts. In process monitoring and control also, little work has been reported. Figure 8.17 summarizes these gaps in an effective and self-explanatory manner.

In this chapter, an attempt has been made to cover all the necessary aspects of extrusion-based AM techniques. The next chapter covers material jetting-based AM processes.

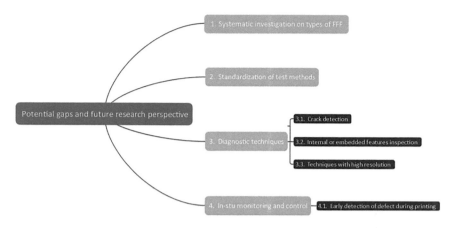

FIGURE 8.17
Potential Gaps and Future Research Perspectives in Field of Extrusion-Based AM Techniques. (*Source:* [4]. Reproduced with permission from Goh et al., Process–structure–properties in polymer additive manufacturing via material extrusion: a review. *Critical Reviews in Solid State and Materials Sciences*, 2019: pp. 1–21.)

References

1. Vaezi, M., Chianrabutra, S., Mellor, B., Yang, S., Multiple material additive manufacturing. Part 1: a review. *Virtual and Physical Prototyping*, 2013. 8(1): pp. 19–50.
2. Gonzalez-Gutierrez, J., Cano, S., Schuschnigg, S., Kukla, C., Sapkota, J., Holzer, C., Additive manufacturing of metallic and ceramic components by the material extrusion of highly-filled polymers: a review and future perspectives. *Materials*, 2018, **11**(5): p. 840.
3. Srivastava, M., *Some studies on layout of generative manufacturing processes for functional components*, 2015, Delhi University, India.
4. Goh, G. D., Yap, Y. L., Tan, H. K. J., Sing, S. L., Goh, G. L., Yeong, W. Y., Process–structure–properties in polymer additive manufacturing via material extrusion: a review. *Critical Reviews in Solid State and Materials Sciences*, 2019: pp. 1–21.
5. Zein, I., Hutmacher, D. W., Tan, K. C., Teoh, S. H., Fused deposition modeling of novel scaffold architectures for tissue engineering applications. *Biomaterials*, 2002, **23**(4): p. 1169–1185.
6. Gibson, I., Rosen, D. W., Stucker, B., Extrusion-based systems, in *Additive manufacturing technologies: rapid prototyping to direct digital manufacturing*, 2010, Springer, pp. 160–186.

9

Additive Manufacturing Processes Utilizing Material Jetting

9.1 Introduction

This is the fourth chapter of Section B of this book, which presents process specific details of various AM processes. In the previous chapter, readers were introduced to the basic principles of the AM processes utilizing extrusion-based systems. This chapter provides all the basic details of the AM processes utilizing material jetting in terms of multi-jet printing basic principle; droplet formation techniques including continuous stream and drop-on-demand ink jet technology; materials for material jetting; advantages and limitations of material jetting; applications; and design and quality aspects. This chapter concludes the discussion with a summary.

Two-dimensional inkjet printing (IJP) has long been used to print colored or black and white documents, as well as images from their digital inputs. This technique has extended itself to the desktop industry and there are a variety of market leaders in this domain, for example HP, Canon, etc. While 2D inkjet printing came into being in the 1960s, 3D IJP emerged around two decades later basically for rapid prototyping requirements.

According to ISO/ASTM [1]: "Material jetting (MJ) is an AM technique involving selective deposition of build material droplets." Bill Master [2] was amongst the pioneers who filed patents in AM technology and founded Perception Systems, Inc. that later became BPM Technology, a company that innovated ballistic particle manufacturing technology. In 1994, the first commercialized machine was made by a company later named Solidscape. In 1996, multi-jet printing was commercialized under the trade names of MJM and Thermojet. In 1998, Polyjet technology was developed by Objet [3] (an Israel-based company which later merged with Stratasys in 2012).

9.2 Variants of Material Jetting

Today, material jetting modellers are provided by various AM providers such as Optomec Inc., 3D Systems Corp., Stratasys, Luxexcel, etc. There are different variants which cover different commercial processes including [4]:

- Aerosol Jet®
- Ballistic particle manufacturing (BPM)
- Drop-on-demand (DOD)
- Laser-induced forward transfer (LIFT)
- Liquid metal jetting (LMJ)
- Multi-jet modelling (MJM)
- Multi-jet printing (MJP)
- Nano Metal Jetting© (XJet) (NMJ)
- NanoParticle Jetting™ (XJet) (NPJ)
- Polyjet®
- Printoptical© Technology
- Thermojet printing.

9.3 Multi-Jet Printing

Multi-jet printing technology utilizes piezoelectric technology for material jetting in layer-by-layer fashion onto the build platform. It is an inkjet printing process, in which material droplets from the printing head are initially deposited upon the surface. These are then permitted to solidify to obtain an initial layer. Then, subsequent layer deposition one by one over the previously deposited layers takes place. Finally, the deposited layers are hardened or cured using UV light.

The basic working principle of MJ includes:

- First step involves heating of resin in the range of 30–60°C for optimizing printing viscosity.
- Second step involves travelling of printhead above build platform and corresponding jetting/deposition of a multitude of tiny photopolymer droplets at prescribed locations.
- In the third step, the printhead which has an attached UV light is responsible for curing and solidification of the deposit for creation of part layer.

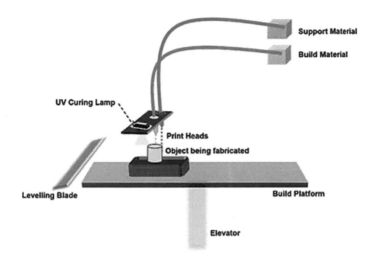

FIGURE 9.1
Schematic Representation of Material Jetting Process. (*Source:* [5]. Sireesha et al., A review on additive manufacturing and its way into the oil and gas industry. *RSC Advances*, 2018, 8(40): pp. 22460–22468, under Creative Commons Licence.)

- In the fourth step, movement of build platform in downward direction by a depth equal to one layer thickness takes place to enable formation of next layer.
- The entire process is repeated till the complete part is fabricated.

A schematic illustration of MJ technology is presented by Figure 9.1.

9.4 Droplet Formation Techniques

The basic idea behind material jetting/inkjet printing is the generation of small droplets of material and their deposition over a substrate to develop a specific pattern. This can be achieved via different techniques. The main modes of inkjet technology used for deposition in MJ include the continuous stream inkjet (CIJ) and drop-on-demand (DOD) techniques. A detailed classification of the main inkjet technologies is presented in Figure 9.2. The main difference between continuous stream and DOD is the timing of droplet generation. In continuous stream, the droplets are generated by breaking up the continuous stream of droplets through an ejection nozzle, whereas in DOD droplets are generated whenever required. Thus, in the

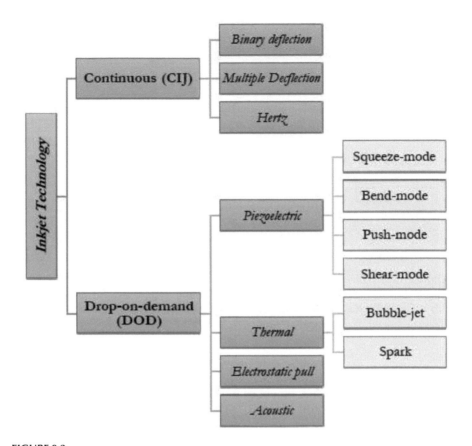

FIGURE 9.2
Overview of Inkjet Technologies with Different Mechanisms of Drop Generation. (*Source:* [8]. Ledesma Fernandez, *Jetting of multiple functional materials by additive manufacturing*, 2018, University of Nottingham, under Creative Commons Licence.)

case of the continuous stream technique ejection of droplet stream and its corresponding recycling via gutter continues even in the non-working state of printer. In contrast, a DOD inkjet printer is supposed to eject an ink droplet only when commanded for the same and not continuous.

MJ offers highest Z axis resolution (minimum 16-micron size layers) amongst all other AM techniques [5–7]. A detailed discussion of these two techniques is presented below.

9.4.1 Continuous Stream Inkjet Technology

Ink droplets are constantly created in this approach. Pressurized liquid pushes a liquid column through a nozzle of small diameter which leads to the creation of droplets. This process is regulated by a high-pressure pumping system which vibrates the nozzle using a piezoelectric crystal. Selective charging of ink drops is accomplished with signals from printer. Deflection of these charged droplets into a gutter occurs during recirculation. On the other hand, ejection of uncharged droplets onto a matrix for image formation takes place [9]. A schematic of CIJ is given in Figure 9.3 showing binary and multiple deflection CIJ systems. Droplets without charge get printed onto a matrix in a binary deflection system which uses only one nozzle for printing at one dot position. However, in multiple deflection systems, several dots are deposited corresponding to each nozzle.

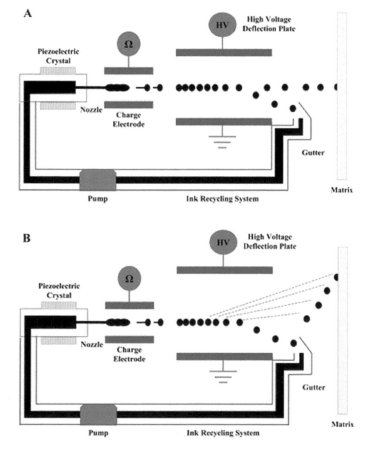

FIGURE 9.3
Schematic of Continuous Inkjet Printing Technology: (a) Binary Deflection System with Single Nozzle; (b) Multiple Deflection System. (*Source:* [9]. Reproduced from Li et al., Inkjet printing for biosensor fabrication: combining chemistry and technology for advanced manufacturing. *Lab on a Chip*, 2015, 15(12), 2538–2558, with permission of the Royal Society of Chemistry.)

In CIJ, the jet breaks naturally at a wavelength characterized by its diameter. This is owing to the fact that surface energy for a sphere is lower than that for a corresponding cylinder of similar volume. Tuning of this breaking phenomenon can be accomplished to achieve specific rates of droplet formation with creation of regular known frequency disturbance with piezoelectric element. For achieving a complex pattern, controlling the position of a droplet train upon substrate as well as selecting pre-specified droplets from the stream is important once the droplet train has initiated. One popular way of controlling these factors is induction of charge into droplets immediately after they are created and their subsequent deflection on flight using an electric field.

9.4.2 Drop-on-Demand Inkjet Technology

DOD inkjet technology is more in demand than CIJ technology [10]. In the DOD approach, individual droplets are generated from the nozzle when required. No additional instrument is needed to select individual drops. This means that elimination of a complicated system for droplet charging, deflecting and recycling makes DOD technology relatively cheaper. A schematic illustration of DOD is presented in Figure 9.4. Finer droplets, approximately of orifice diameter less than 20-micron size, with enhanced accuracy of placement are achieved using DOD. A pressure pulse is utilized for creation of ink droplets in a DOD printer. Generation of a pressure pulse can occur in a variety of ways which define the DOD subclasses. These ways include thermal, piezoelectric, acoustic, electrostatic and electro-hydrodynamic systems. Thermal and piezoelectric methods dominate modern IJP techniques. Electro-hydrodynamic systems have recently been gaining prominence. Other methods are more or less still under development before they can be effectively utilized. A common actuation technique is placement of the piezoelectric element inside or next to the material chamber. Based upon position as well as mechanism of expansion and contraction, systems can be designed as squeeze, bend, push or shear mode as shown in Figure 9.5.

A real-time image of the printing head and platform of the ProJet 5500X multi-material system is shown in Figure 9.6(a), and the working principle of the same modeller is shown in Figure 9.6(b). This system contains many IJP printheads based on the piezoelectric principle and has more than 100 nozzles to accomplish printing of two materials along with support material. Material is cured with a UV laser. A planerizer rolls over the printed layer and is utilized to level and remove excess material for desired layer thickness.

A basic comparison of CIJ and DOD inkjet printing techniques is presented in Table 9.1.

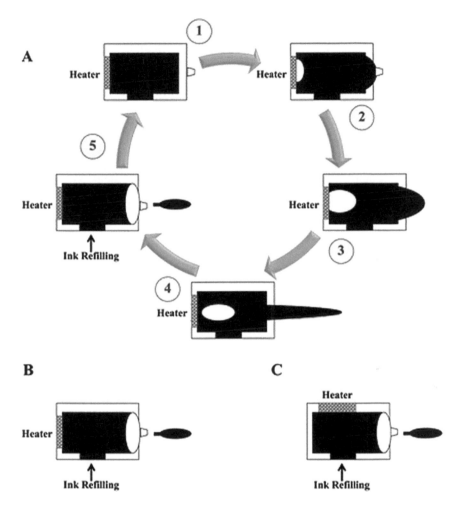

FIGURE 9.4
DOD Thermal Inkjet Printer Showing: (A) Mechanism; (B) Thermal "Roof-Shooter"; (C) "Side-Shooter" Configuration. (*Source:* [9]. Reproduced from Li et al., Inkjet printing for biosensor fabrication: combining chemistry and technology for advanced manufacturing, *Lab on a Chip*, 2015, 15(12), 2538–2558, with permission of the Royal Society of Chemistry.)

9.5 Materials for Material Jetting

Material jetting is used to print/deposit materials such as polymers (for example ABS, polyamide, PLA, etc.), metals, ceramics and polymer composites. In addition to individual materials, a combination of different types of materials can be printed using MJ.

TABLE: 9.1

Comparison between CIJ and DOD Inkjet Printing Techniques

CIJ	DOD
Drop generation is constant/continuous	Drop generation is as per demand/
High-speed printing and is useful for industrial	requirement
applications	Comparatively less printing speed
Problem of clogging of nozzle is less, owing to	Cost in DOD systems is relatively less
continuous flow of droplets	owing to elimination of deflection and
Lower resolution owing to high speed	recycling system and complex droplet
CIJ printers are relatively costly owing to	charging
requirements of drop selection, recycling	Maintenance cost is comparatively less
system	Almost all commercial material jetting
Maintenance cost is high	machines utilize DOD printheads

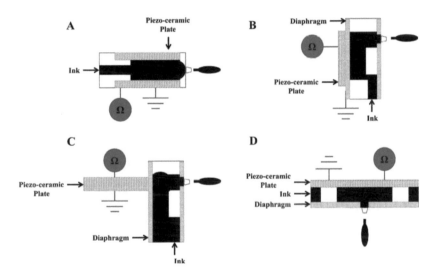

FIGURE 9.5
DOD Piezoelectric Inkjet Printer: (A) Squeeze Mode; (B) Bend Mode; (C) Push Mode; (D) Shear Mode. (*Source:* [9]. Reproduced from Li et al., Inkjet printing for biosensor fabrication: combining chemistry and technology for advanced manufacturing, *Lab on a Chip*, 2015, 15(12), 2538–2558, with permission of the Royal Society of Chemistry.)

9.6 Advantages, Drawbacks and Applications of Material Jetting

9.6.1 Advantages of MJ

One of the biggest advantages of this technology is the possibility of utiliz-ing more than one printing head simultaneously, which gives them the

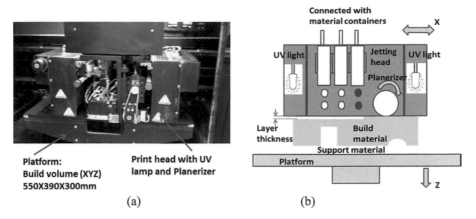

Platform:
Build volume (XYZ)
550X390X300mm

Print head with UV
lamp and Planerizer

(a)

(b)

FIGURE 9.6
Multi-Material IJP (ProJet 5500X) System: (a) Platform and Printhead; (b) Working Principle.
(*Source:* [11]. Reproduced with permission from Yang et al., Performance evaluation of ProJet
multi-material jetting 3D printer. *Virtual and Physical Prototyping*, 2017, 12(1): pp. 95–103.)

ability to print special parts, for example, fast printing by covering the build
surface entirely in a single pass and printing multiple materials simultaneously. Additionally, appreciably better surface qualities are obtained owing
to quite small drops being jetted. Some common advantages of MJ are summarized below:

- Easy operation and safe to handle owing to absence of loose powder
- Good surface finish
- No need for post-curing
- Components fabricated via MJ have homogeneous mechanical and
  thermal properties.

9.6.2 Drawbacks of MJ

Since an appreciably small amount of materials (in the form of small droplets) is jetted over a small build area, build rates are generally less. The
main drawbacks of this technology are:

- Build volume is small
- Material cost is higher
- Support structure is needed
- Time to print/develop newer materials is quite long.

9.6.3 Applications of MJ

There are numerous applications of the MJ process. A few noteworthy applications are in the fabrication of:

- Light honeycomb structures
- Scaffolds for tissue engineering
- Custom anatomical models
- Biosensors
- IJPs conductive circuit traces, LCD color filters and plasma displays.

A few illustrative applications of MJ from relevant research are presented in Figures 9.7 and 9.8.

(a)

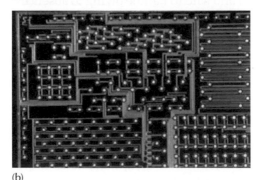

(b)

FIGURE 9.7

3D Printed Objects: (a) Inkjet Printed Ceramic Object (Impeller); (b) Solder Droplets Printed IC Test Board. (*Source:* (a) [12]. Reprinted Ainsley et al., Freeform fabrication by controlled droplet deposition of powder filled melts. *Journal of Materials Science*, 2002, 37(15): pp. 3155–3161, Springer, with permission from Springer Nature. (b) [13]. Reprinted from Liu and Orme, High precision solder droplet printing technology and the state-of-the-art. *Journal of Materials Processing Technology*, 115(3): pp. 271–283, Copyright 2001, with permission from Elsevier.)

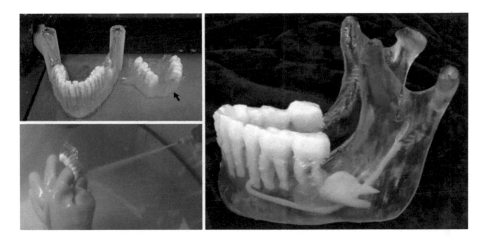

FIGURE 9.8
Printed Model of a Mandible via Material Jetting Printer. *Note:* Support material (black arrow in top left image shows support material) is removed using water jet and allowed to dry. Right panel shows final product having internal features (teeth, impacted molar, alveolar canal and cyst highlighted. *(Source:* [14]. Reprinted/adapted from Mitsouras and Liacouras, 3D printing technologies, in *3D printing in medicine: a practical guide for medical professionals,* F. J. Rybicki and G. T. Grant, Editors, 2017, Springer International, pp. 5–22, with permission from Springer Nature.)

9.7 Design and Quality Aspects

Based upon discussion in the preceding sections, it can be concluded that MJ has currently secured the stature of a successful and prominent AM technique. However, there are some challenges which are restricting the growth of MJ. These mainly include [15]:

- *Liquid material formulation:* Suspension of particles in carrier, dissolution into solvent, melting of polymer, mixing a monomer/ prepolymer formulation with PI molecule, etc. are the steps that need to be accomplished if the initial raw material state is not liquid. Sometimes, additives like surfactants need to be added for desirable features. IJP inks for AM should also follow trends of two-dimensional printing inks where there are innumerable industries solely devoted to ink manufacturing.

- *Droplet formation:* The second challenge is to obtain small discrete droplets from the liquid material. The droplet formation is largely dependent on the optimum relationship between process parameters, hardware involved, type of material being printed, and method of droplet formation. Change in material, such as adding tiny particles, changes its physical set-up and droplet forming behavior.

• *Deposition of droplets:* This challenge relates to control of droplet deposition which mainly includes substrate wetting, impact and flight path of droplets. During MJ, either build platform or printhead is normally moving, which needs to be fully considered during droplet trajectory planning. Additionally, deposition traits are also affected by droplet size and velocity. This in turn can be monitored by controlling design and operating principle of nozzle.

9.8 Summary

Each manufacturing technique has some pros and cons. Owing to simplicity, high-speed, economical, enhanced resolution as well as brilliant part characteristics, IJP is gaining edge as a product fabrication tool for varied parts possessing different degrees of intricacy and sophistication. Currently, MJ is the only technology that offers voxel level material property and color tuning.

MJ machines utilize mainly two droplet mechanisms, namely CIJ and DOD. However, nowadays, almost all material jetting machines use DOD mechanisms. Despite the growth of AM related to material jetting technologies, several aspects in this field are still unexplored.

In this chapter, an attempt has been made to cover all the necessary aspects of material jetting-based AM techniques. The next chapter covers the binder jetting-based AM processes.

References

1. International Organization for Standardization (ISO) *ISO/ASTM 52900: Additive manufacturing — General principles — Terminology*, 2015.
2. Masters, W. E., *Computer automated manufacturing process and system*, U.S. Patent, 4665492, May 12, 1987.
3. Gothait, H., *Apparatus and method for three dimensional model printing*, U.S. Patent 6259962, July 10, 2001.
4. Silbernagel, C. *Additive Manufacturing 101–4: What is material jetting?* Available from http://canadamakes.ca/what-is-material-jetting, accessed March 23, 2019.
5. Sireesha, M., Lee, J., Kranthi, K. A. S., Babu, V. J., Kee, B. B. T., Ramakrishna, S., A review on additive manufacturing and its way into the oil and gas industry. *RSC Advances*, 2018, **8**(40): pp. 22460–22468.
6. Sirringhaus, H., Shimoda, T., *Inkjet printing of functional materials. MRS Bulletin*, 2003, **28**(11): pp. 802–806.

7. Ebert, J., Özkol, E., Zeichner, A., Uibel, K., Weiss, Ö., Koops, U., Telle, R., Fischer, H., Direct inkjet printing of dental prostheses made of zirconia. *Journal of Dental Research*, 2009, **88**(7): pp. 673–676.

8. Ledesma Fernandez, J., *Jetting of multiple functional materials by additive manufacturing*, 2018, University of Nottingham.

9. Li, J., Rossignol, F., Macdonald, J., *Inkjet printing for biosensor fabrication: combining chemistry and technology for advanced manufacturing. Lab on a Chip*, 2015, **15**(12): pp. 2538–2558.

10. Kholghi Eshkalak, S., Chinnappan, A., Jayathilaka, W. A. D. M., Khatibzadeh, M., Kowsari, E., Ramakrishna, S., A review on inkjet printing of CNT composites for smart applications. *Applied Materials Today*, 2017, **9**: pp. 372–386.

11. Yang, H., Lim, J. C., Liu, Y., Qi, X., Yap, Yee L., Dikshit, V., Yeong, Wai Y., Wei, J., Performance evaluation of ProJet multi-material jetting 3D printer. *Virtual and Physical Prototyping*, 2017, **12**(1): pp. 95–103.

12. Ainsley, C., Reis, N., Derby, B., Freeform fabrication by controlled droplet deposition of powder filled melts. *Journal of Materials Science*, 2002, **37**(15): pp. 3155–3161.

13. Liu, Q., Orme, M., High precision solder droplet printing technology and the state-of-the-art. *Journal of Materials Processing Technology*, 2001, **115**(3): pp. 271–283.

14. Mitsouras, D., Liacouras, P. C., 3D printing technologies, in *3D printing in medicine: a practical guide for medical professionals*, F. J. Rybicki and G. T. Grant, Editors, 2017, Springer, pp. 5–22.

15. Gibson, I., Rosen, D., Stucker, B., Material jetting, in *Additive manufacturing technologies: 3D printing, rapid prototyping, and direct digital manufacturing*, 2015, Springer, pp. 175–203.

10

Additive Manufacturing Processes Utilizing Binder Jetting

10.1 Introduction

This is the fifth chapter of Section B of this book, which presents process specific details of various AM processes. In the previous chapter, readers were introduced to all the basic details of the AM processes utilizing material jetting-based AM techniques. This chapter covers AM processes utilizing binder jetting (BJ) with respect to various aspects, including: process description; raw materials; design and quality aspects of BJ; advantages and limitations of BJ; and applications. This chapter concludes the discussion with a summary.

BJ was originally developed by Michal J. Cima and co-workers at Massachusetts Institute of Technology in 1989 with the name of "3D printing" (3DP), with its subsequent licensing to various companies around 1993 which includes ZCorp (1994, later taken over by 3D Systems around 2011) and Ex One (1996). A new BJ technique was introduced by HP around 2014 which utilized an integrated heating device. It was claimed that a speed enhancement of about ten times over competitive AM techniques like laser sintering was obtained using this principle. This technique became so popular that AM techniques have long been used synonymously with 3D printing, which is actually a misnomer since 3D printing is one special BJ process. BJ can also be known [1] as:

- three-dimensional printing [2]
- powder bed/inkjet head 3D printing (PBIH) [3]
- inkjet printing (IJP) [4]
- ColorJet Printing (CJP) (3D Systems Corp.)
- Plaster-based 3D printing (PPP) [3]
- MultiJet Fusion™ (Hewlett-Packard Development Company)
- Digital Metal® (Höganäs AB)
- Z printing [5].

In contrast to other printing processes such as material jetting (described in Chapter 9), BJ works on the principle of deposition of liquid binder at pre-specified areas as per 3D CAD model. Thus, BJ is an AM process in which selective deposition of liquid binder over a powder surface takes place resulting in joining of powder particles. The liquid binding agent in the form of droplets is deposited through the printhead followed by the lowering of the platform to add subsequent powder layer and binding agent. This is repeated till the desired build dimensions are obtained.

10.2 Process Description

The BJ process is identical to two-dimensional printing in principle. However, as compared to printing on paper sheet as is done in 2D printing, 3D objects are printed as an output to this process. A thin layer of material in loose powdered form (0.05 to 0.5 millimetres) is first spread upon the build substrate. This is followed by back and forth movement of a traditional inkjet printing head over the previously spread powder and consequent deposition of a binding material, in contrast to the ink used in 2D printing upon areas that need to be transformed into a solid part. A process schematic of BJ is shown in Figure 10.1. The binding agent can be transparent or colored in nature and is in a liquid state initially. Binding of powdered material normally involves a chemical reaction which is dependent upon method/materials utilized. This may occur under the following

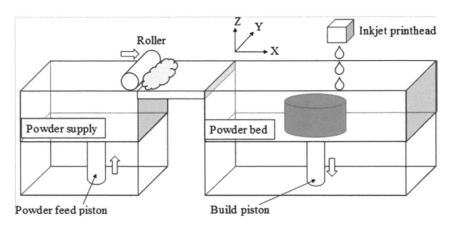

FIGURE 10.1
Binder Jetting 3D Printing Technology Process Schematic. (*Source:* [6]. Reprinted Zhang et al., Additive manufacturing of metallic materials: a review. *Journal of Materials Engineering and Performance*, 2017, 27(1): pp. 1–13, Springer, with permission from Springer Nature.

four situations: (1) when binder and raw powder are in contact; (2) when evaporation of binder occurs owing to its contact with air; (3) when binder is in contact with another chemical mixed with powder; and (4) during activation of binder using heat. The build platform is then lowered in height for deposition of next powder layer and the entire process is repeated till the desired part is obtained. Layer mechanical strengthening due to partial evaporation of water content is needed for ensuring feature structural integrity as well as the required dimensional accuracy. These green parts possess low strength and need to be infiltered/sintered during post-processing in order to obtain the desired mechanical strength.

The basic procedure of the BJ process is summarized below:

- Initial spreading of powder materials upon platform
- Deposition of binder adhesive upon powder top by printhead at required places
- Lowering of build platform and spreading next layer upon previously deposited layer takes place. Formation of object takes place at points of binding of powder with liquid.
- These steps are repeated till the entire object has been made.

The common steps of the BJ process are illustrated in Figure 10.2.

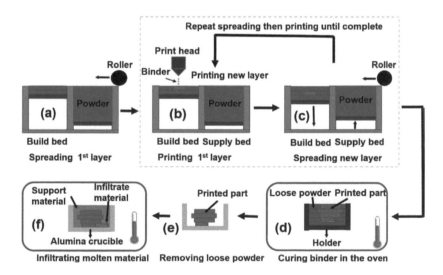

FIGURE 10.2
Steps in Binder Jetting Additive Manufacturing. (*Source:* [7]. Reprinted with permission from Do et al., Process development toward full-density stainless steel parts with binder jetting printing. *International Journal of Machine Tools and Manufacture*, 121: pp. 50–60, Copyright 2017.

Printheads (PH) resemble those used in 2D printing, i.e. piezoelectric or thermal in nature. Binders are supposed to exhibit traits similar to original ink. The binder should preferably possess lower viscosity (in tens of centipoise range) to enable its jetting via fine nozzles present in the printhead. Piezoelectric PHs possess a piezoelectric mechanism connected to a diaphragm that can push material from the nozzle as drops. Thermal or bubble jet PHs contain a thermal heating element which initiates boiling of the binder leading to bubble formation in PH chamber resulting in it being pushed from the nozzle and subsequent droplet formation. Advanced BJ systems by HP utilize application of detailing agent upon a heat-activated binding agent leading to neutralization of any probable binding which results in clearly separated fused and unfused particles, thereby exhibiting enhanced surface finish. Also, the heat-activated binding agent results in parts with better material strength as compared to conventional BJ processes.

10.3 Raw Materials

Polymers (such as ABS, PA, PC), metals (such as Al, stainless steel, etc.), sands and ceramics can be successfully processed using BJ. BJ is one of the most suitable AM process for fabricating ceramic and refractory metal parts which are generally difficult to manufacture using other AM techniques.

10.4 Design and Quality Aspects of BJ

During the BJ process, there are several factors which affect the properties of the final prepared product. These factors may be roughly divided into two categories: process and material related factors/parameters [8]; they are discussed below in detail.

10.4.1 Process Related Parameters

The parameters related to the BJ process which affect the quality of fabricated parts mainly include spreading speed, binder saturation, in-process heating, thickness of layers, etc. and are discussed below:

- *Spreading speed:* Spreading speed is basically the rate at which the rotating roller traverses in a forward direction to spread a layer of powder. In other words, it can be understood as the rate at which a single layer of powder is spread. Generally, a low spreading speed is desirable for uniform powder spreading; however, with lower spreading speed, printing time increases [8]. Thus, the optimum spreading speed should be chosen in order to obtain sound components.

- *Binder saturation level:* The binder saturation level can be defined as the amount of binder deposited via the printhead. At low binder saturation levels, the amount of binder is insufficient to bind powder particles together and this results in poor bonding. On the other hand, at higher saturation levels, the excess binder permeates out of the defined boundaries. An example showing the effect of variation of binder saturation level by keeping other parameters, such as particle size and size distribution, constant is presented in Figure 10.3. It shows a 3D printed mesh structure of TiNiHf with three settings of saturation level. The printed structure with 55% saturation level is extremely fragile and breaks down when it is transferred from printbed to part containing station showing that the selected saturation level is not enough for fabrication of sound 3D parts (see Figure 10.3(a)). Figure 10.3(b) and (c) show mesh structures at 110% and 170% saturations respectively. The mesh structure with 110% saturation level is better than the structure at 55%, but is still weak. The structure obtained at saturation level of 170% exhibits enough strength to be removed from the printbed. For obtaining dimensional accuracy, higher mechanical and structural properties of printed parts, the optimum binder level should be selected.

- *In-process heating:* During BJ process, the deposited binder on a particular layer is partially cured using an external heat source (usually radiation heat source, e.g. heating lamp) before spreading the subsequent layer's powder. This heating may result in reduction of

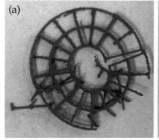 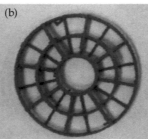 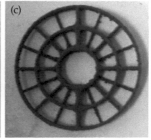

FIGURE 10.3
TiNiHf Mesh Structures Having Printing Layer Thickness of 50 μm but Varying Binder Saturations: (a) 55%; (b) 110%; (c) 170%. (*Source:* [9]. Reprinted from Lu and Reynolds, 3DP process for fine mesh structure printing. *Powder Technology*, 187(1):, pp. 11–18, Copyright 2008, with permission from Elsevier.)

binder saturation level via evaporation [8]. Thus, an optimum in-process heating is applied which can be controlled by altering the intensity of applied heat and drying time.

- *Thickness of layers:* Layer thickness is generally determined by designers/operator and is fed in the software to estimate the binder saturation level. With the change in layer thickness the amount of binder saturation level also changes. Generally, higher layer thickness produces lower resolution [8] but better powder spreading [9] and vice versa. Thus, layer thickness affects the binder saturation level and consequently dimensional accuracy, mechanical characteristics and green part resolution. A typical example (discussed above, in case of binder saturation level) showing the effect of variation of layer thickness is shown in Figure 10.4 which describes the effect of layer thickness on breaking strength of the 3D printed mesh structure.

10.4.2 Material Related Parameters

Material related parameters that affect the finished product characteristics can be divided into powder related properties (such as particle size and distribution) and binder related properties (such as viscosity and surface tension of binder) and are discussed below [8]:

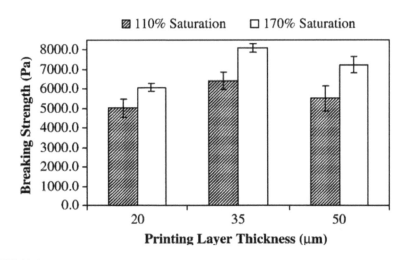

FIGURE 10.4
Effect of Layer Thickness and Saturation Level on Breaking Strength of TiNiHf 3D Mesh Structures. (*Source:* [9]. Reprinted from Lu and Reynolds, 3DP process for fine mesh structure printing. *Powder Technology*, 187(1):, pp. 11–18, Copyright 2008, with permission from Elsevier.)

- *Powder particle size:* Powder particle size directly affects flow-ability of powder. As per Teunou et al. [10] large powder particles facilitate easy flow while fine particles tend to agglomerate owing to inter-particle forces. Thus, an optimum particle size of powder should be selected for desired density of the component.

- *Shape of powder particles:* Similar to particle size, the shape of powder particle has great importance in determining the characteristics of manufactured product. The shape of particles affects the shrinkage during sintering, packing density and flowability to some extent. Results of an experimental research study showing the effect of particle morphology over shrinkage are presented in Figure 10.5. It shows microstructural images of two different morphologies of 316-stainless steel powders processed via water and gas atomization and the effect of these shapes on shrinkage during sintering. It can be easily estimated from the figure that the particles with irregular shapes causes the highest shrinkage.

- *Viscosity and surface tension of binders:* During the BJ technique, binder droplets are deposited on the surface of the powder and then these droplets migrate into the powder bed by capillary action with different rates in all directions [8]. Owing to the different migration rates, liquid phase under equilibrium assumes particular profiles that depend upon the physical characteristics of liquid binder like viscosity and surface tension, as also interaction between binder and powder bed which further influences the green strength of the

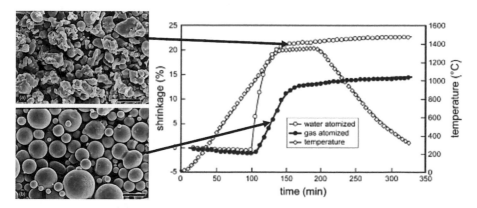

FIGURE 10.5
Power Particle Morphology's Effect on Shrinkage of Stainless Steel during Sintering (Conventional Furnace). (*Source:* [11]. Reprinted from Koseski et al., Microstructural evolution of injection molded gas- and water-atomized 316L stainless steel powder during sintering. *Materials Science and Engineering: A*, 390(1):, pp. 171–177, Copyright 2005, with permission from Elsevier.)

fabricated part. Thus, the physical properties of binders affect the quality of manufactured parts via the BJ process.

Apart from the above discussed factors, other aspects that should be given weightage while designing the BJ process mainly include considerations related to: powder feedstock, printhead, and powder bed–binder interaction [12]. The important points of these considerations are briefly summarized below:

- *Powder feedstock considerations:* During the BJ process the fabrication of the part is realized by selectively applying binder upon the powder bed that can be understood as a classic liquid–porous media interaction case. The interaction between binder and powder largely depends on the characteristics of these two types of materials. The individual effects of powder and binder characteristics, such as particle size and shape, binder physical properties, etc. are described above. In addition, the powder spreading mechanisms (which mainly include forward and backward rolling rollers and blades) also affect the powder bed densities. According to various experimental research studies [13, 14], higher powder density is achieved using a backward rolling roller mechanism. Another noteworthy aspect in feedstock consideration is the minimum feature resolution, which is determined by several factors such as mean particle size, layer thickness, etc.

- *Printhead considerations:* Generally, BJ-AM systems utilize a drop-on-demand (DOD) printhead. A detailed discussion explaining the DOD mechanism has already been presented in Chapter 9. For a particular binder system, the precise and rapid delivery of droplets with control volume strongly depends on optimized printhead control. The PHs of some BJ systems can deliver droplets in adjustable/fixed quantities. Binder saturation (ratio of binder liquid volume to that of all pores/voids within same volume) monitoring is related to the ability of the PH to deliver variable liquid quantities to specified positions. For constant volume droplet PHs, minimal saturation level resolutions greatly depend on single drop volume, as shown in Figure 10.6, and can be adjusted by either overlapping or overlaying droplets, as illustrated in Figures 10.6(a), (b) and (c) respectively. Overlapping permits saturation level control with extremely fine "step size" that can be accounted to continuous adjustability of overlapping ratio. Contrarily, overlaying amounts to high saturation levels in narrow and deep regions and is therefore suitable for printing thicker layers.

  Another important aspect is the printing speed of the process. It is always desirable to achieve high printing speed for enhancing the production rates. In BJ systems, this can be achieved by enhancing printhead moving velocity and shortening in-process drying time. The effect of printing and spreading speed has been discussed in detail above. The other aspect demanding consideration is clogging of the printhead. Clogging in BJ is basically the state when there is temporary clogging of the nozzle during BJ printing which results in missing print lines and disintegrated parts. An example is illustrated in Figure 10.7.

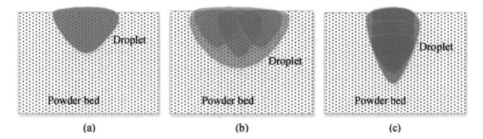

FIGURE 10.6
Control Methods of Saturation Level: (a) Individual Droplet; (b) Overlap; (c) Overlay. (*Source:* [12]. Reprinted from Miyanaji et al., Process development for green part printing using binder jetting additive manufacturing. *Frontiers of Mechanical Engineering*, 2018, 13(4): pp. 504–512, Springer, with permission from Springer Nature.

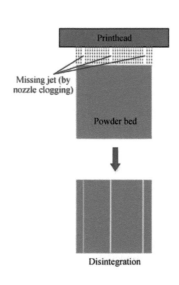

FIGURE 10.7
Printed Defects owing to Nozzle Clogging. (*Source:* [12]. Reprinted from Miyanaji et al., Process development for green part printing using binder jetting additive manufacturing. *Frontiers of Mechanical Engineering*, 2018, 13(4): pp. 504–512, Springer, with permission from Springer Nature.

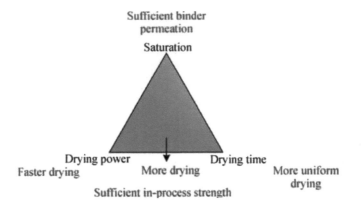

FIGURE 10.8
In-Process Printing Parameters. (*Source:* [12]. Reprinted from Miyanaji et al., Process development for green part printing using binder jetting additive manufacturing. *Frontiers of Mechanical Engineering*, 2018, 13(4): pp. 504–512, Springer, with permission from Springer Nature.

- *Consideration related to powder bed–binder interaction:* During the BJ process, the properties of fabricated parts is largely dependent on the interaction of powder bed and binder material. As the binder falls over the powder, it has some impact speed which dissipates quickly. Once it starts migrating in the powder, the behavior of the result depends on the characteristics of the binder and powder material. Sufficient wetting of the powder and binder is of utmost importance in defining the soundness of the manufactured product. The in-process drying and level of binder saturation are also important factors which should be adequately adjusted for achieving better properties of the resultant component. Figure 10.8 presents the relationship between saturation and permeation level, drying time and process strength during the BJ process.

10.5 Advantages of Binder Jetting

The BJ process shares some common benefits with material jetting (MJ). However, it has some distinct advantages also as compared to MJ which are summarized in points below:

- Ability to print colored components using colored binders
- Wide range of material compatibility
- Process speed is high
- Lower heat related defects

- Larger build volumes
- No need for support structure
- Low cost per part owing to inexpensive binders and materials
- Negligible or no residual stress.

10.6 Drawbacks of Binder Jetting

In addition to the major advantages of high process speed, larger build volumes, etc., BJ also has some drawbacks, which are summarized below:

- Part strength is less, especially where no external heat is utilized for binding of parts
- Not suitable for structural parts
- Multiple post-processing steps are needed
- Disturbances during depowdering (taking part out from powder bed build area)
- Formation of porosity defects.

10.7 Applications of Binder Jetting

BJ has great applications in printing of components for key industries such as aerospace, automobile, foundry, biomedicine, jewellery, etc. Some examples of BJ fabricated components are shown in Figure 10.9.

FIGURE 10.9
Typical Examples of Components Fabricated Using Binder Jetting: (a) Bevel Gears; (b) Casting Molds for Gear Manufacturing; (c) Al Alloy Gear Developed Using the Mold. (*Source:* [15]. Budzik, G., The use of the rapid prototyping method for the manufacture and examination of gear wheels, in *Advanced applications of rapid prototyping technology in modern engineering*, M. Enamul Hoque, Editor, 2011, InTechopen, under Creative Commons Licence.)

10.8 Summary

The BJ process has many unique advantages as compared to other AM techniques. In comparison with the MJ process, BJ has the unique benefits of high printing speeds, and ability to print colored components using colored binders. Various factors, including binder related, powder related, feedstock, process hardware, etc. should be selected at their optimum values to achieve better properties of manufactured components.

In this chapter, an attempt has been made to cover all the necessary aspects of binder jetting-based AM techniques. The next chapter covers sheet lamination-based AM processes.

References

1. Silbernagel, C. *Additive manufacturing 101–1: What is binder jetting?* Available from: http://canadamakes.ca/what-is-binder-jetting, accessed February 8, 2019.
2. Sachs, E. M., Haggerty, J. S., Cima, M. J., Williams, P. A., *Three-dimensional printing techniques*, U.S. Patent 5204055, April 20, 1993.
3. Cotteleer, M., Holdowsky, J., Mahto, M., *The 3D opportunity primer: the basics of additive manufacturing*, 2014, Deloitte University Press, Westlake, Texas. Available from: http://dupress.com/articles/the-3d-opportunity-primer-the-basics-of-additive-manufacturing, accessed February 8, 2019.
4. Sachs, E., Cima, M., Williams, P., Brancazio, D., Cornie, J., Three dimensional printing: rapid tooling and prototypes directly from a CAD model. *Journal of Engineering for Industry*, 1992, **114**(4): pp. 481–488.
5. Iliescu M., Nutu, E., Tabeshfar, K., Ispas, C., Z printing rapid prototyping technique and solidworks simulation: major tools in new product design, in *Proceedings of the 2nd WSEAS International Conference on Sensors, and Signals and Visualization, Imaging and Simulation and Materials Science*, 2009, World Scientific and Engineering Academy and Society (WSEAS), Baltimore, USA: pp. 148–153.
6. Zhang, Y., Wu, L., Guo, X., Kane, S., Deng, Y., Jung, Y-G., Lee, J-H., Zhang, J., Additive manufacturing of metallic materials: a review. *Journal of Materials Engineering and Performance*, 2017, **27**(1): pp. 1–13.
7. Do, T., Kwon, P., Shin, C. S., Process development toward full-density stainless steel parts with binder jetting printing. *International Journal of Machine Tools and Manufacture*, 2017, **121**: pp. 50–60.
8. Miyanaji, H., *Binder jetting additive manufacturing process fundamentals and the resultant influences on part quality*, 2018, Department of Industrial Engineering, University of Louisville, Louisville, Kentucky.
9. Lu, K., Reynolds, W. T., 3DP process for fine mesh structure printing. *Powder Technology*, 2008, **187**(1): pp. 11–18.
10. Teunou, E., Fitzpatrick, J. J., Synnott, E. C., Characterisation of food powder flowability. *Journal of Food Engineering*, 1999, **39**(1): pp. 31–37.

11. Koseski, R. P., Suri, P., Earhardt, N. B., German, R. M., Kwon, Y.-S., Microstructural evolution of injection molded gas- and water-atomized 316L stainless steel powder during sintering. *Materials Science and Engineering: A*, 2005, **390**(1): pp. 171–177.
12. Miyanaji, H., Orth, M., Akbar, J. M., Yang, L., Process development for green part printing using binder jetting additive manufacturing. *Frontiers of Mechanical Engineering*, 2018, **13**(4): pp. 504–512.
13. Budding, A., Vaneker, T. H. J., New strategies for powder compaction in powder-based rapid prototyping techniques. *Procedia CIRP*, 2013, **6**: pp. 527–532.
14. Shanjani Y, Toyserkani, E. Material spreading and compaction in powder-based solid freeform fabrication methods: mathematical modeling, in *Proceedings of International Solid Freeform Fabrication (SFF) Symposium*, 2008, Austin.
15. Budzik, G., The use of the rapid prototyping method for the manufacture and examination of gear wheels, in *Advanced applications of rapid prototyping technology in modern engineering*, M. Enamul Hoque, Editor, 2011, InTechopen.

11

Additive Manufacturing Processes Utilizing Sheet Lamination Processes

11.1 Introduction

This is the sixth chapter of Section B of this book, which presents process specific details of various AM processes. In the previous chapter, readers were introduced to all the basic details of the AM processes utilizing binder jetting-based AM techniques. This chapter deals with AM processes utilizing sheet lamination processes in terms of its variants; laminated object manufacturing including its process description, materials, process variants, advantages, limitations and applications; ultrasonic consolidation including its basic principles, advantages, limitations and applications; and design and quality aspects. This chapter concludes the discussion with a summary.

According to the ISO/ASTM definition, "Sheet lamination is AM process in which sheets of material are bonded to form a part." Sheet lamination processes (SLPs) sequentially stack, laminate and shape thin material sheets to develop 3D components. These are best understood as hybrid processes which utilize material addition as well as subtraction principles. In these techniques, layers are laminated (additive principle) to build the part and then they are cut using a tool such as laser beam, knife, CNC machining, etc. (subtractive process). Each subsequent layer is bonded with the previous layer using different means based on particular processes.

11.2 Variants of Sheet Lamination

Sheet lamination has mainly two popular variants, i.e. laminated object manufacturing (LOM) and ultrasonic consolidation (UC). However, there are few other variants that include [1]:

- Computer aided manufacturing of laminated engineering materials (CAM-LEM)

- Plastic sheet lamination (PSL)
- Selective deposition lamination (SDL).

The first LOM machine was commercialized by the Helisys company (1991), which was succeeded by Cubic Technologies. Solidimension (Israel, 1999) introduced a similar system which utilized PVC plastic sheets instead of paper (as in LOM). In 1999, Solidica (now Fabrisonic, a USA-based company) invented a new hybrid process, named ultrasonic consolidation/ultrasonic additive manufacturing, resembling LOM. However, they used metal tapes that were additively joined by ultrasonic vibrations. These were subsequently machined using subtractive milling operations (such as CNC) to provide the desired shape and dimensions. In 2005, Kira (a Japanese company) introduced a machine similar to LOM which used a steel cutter instead of a laser beam. Mcor Technologies (2008) launched its pioneer SDL modeller under the name of Matrix that was similar to LOM. However, it utilized separate A4 paper sheets instead of paper roll which were finally cut using a steel cutter.

11.3 Laminated Object Manufacturing

Kunieda (1984) initially created the concept of LOM [2] and subsequently in 1986 it was further developed by a company named Helisys [3]. In 1991 the company commercialized the LOM machine, which made 3D products with paper rolls and a CO_2 laser. The raw material used in LOM was copier paper which is coated with thermally activated adhesive on one side. The bonding and lamination of stacking layers is achieved using real-time heating and compression by a heated roller. When a heated roller passes over the stacked sheets, the coating on one side of the sheet/paper melts and forms the bond. The build stack is cut by a CO_2 laser to provide the desired 3D shape. During the LOM process, smoke and localized flame are generated and a chimney is often built into the system to encounter these issues.

11.3.1 Process Description

A feed roll is used to enable movement of a layer sheet material adhesively coated from one side in the build area. This sheet is placed over the substrate keeping the adhesive side down. Then, a heated roller passes over the layer and pressure is exerted which results in melting of the adhesive layer and bonding between the substrate and sheet. Subsequently, a laser beam (generally CO_2) traces the outline of one slice of the part. The extra material that does not form the part is then crosshatched by laser. This is followed by the lowering of the platform by a distance which is equivalent to deposited layer thickness. Another sheet is then stacked on the previous layer and the roller is moved on to provide bonding between the stacking sheets. A laser beam directed by a mirror and optical head is utilized for cutting the stacked build for providing the desired shape and dimensions. Then again, the build platform is lowered down and the same process is repeated up to the desired build height. The excess sheet material is hatched and removed using a laser beam and collected from a waste take-up roll. The final component is then obtained. The LOM technique is schematically presented in Figure 11.1.

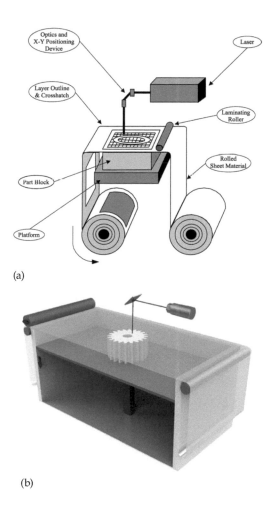

(a)

(b)

FIGURE 11.1

(a) Schematic of Laminated Object Manufacturing; (b) Laminated Object Manufacturing. (*Source: (a)* [4]. Upcraft and Fletcher, The rapid prototyping technologies. *Assembly Automation,* 2003, 23(4): pp. 318–330. Copyright Emerald Publishing Limited, all rights reserved; *(b)* [5]. Chen et al., 3D printing of ceramics: a review. *Journal of the European Ceramic Society,* 2019, 39(4): pp. 661–687, under Creative Commons Licence.)

11.3.2 Materials for LOM

Thin sheets of paper, plastics, metals, ceramics, cellulose, polycarbonate composites, etc. can be used as raw materials for LOM.

11.3.3 Process Variants of LOM

Several process variants to LOM are available, such as Kira's Paper Lamination Technology which utilizes a knife instead of a laser beam to cut each subsequent layer. Solido Ltd. (Israel) also utilizes a knife, but solvent is used instead of adhesive. Some other variants are hybrid additive- and subtractive-based, such as Thick Layer Lamination from Stratoconception (France), Adaptive-Layer Lamination developed by Landfoam Topographics, etc. Helisys, which was the commercial giant of laser-based LOM modellers, ceased operations in 2000. It has now been succeeded by Cubic Technologies.

11.3.4 Advantages of LOM

LOM offers several advantages over other AM techniques which include:

- No support structure is needed
- High manufacturing speed
- Material cost is low
- LOM set-up cost is moderate
- Low thermal stresses
- Low distortion and deformation during processing
- No need for chemical reaction
- Large components can be easily made.

11.3.5 Drawbacks of LOM

Despite the quick operation and cheap raw material, LOM has not been widely established as a prominent AM technology. The main drawbacks of LOM include:

- Poor interfacial bonding between layers
- Only suitable for laminated sheets
- Poor surface finish
- Dimensional accuracy is less
- Difficulty in producing hollow parts

- Waste material is not reusable, but can be recycled
- Skilled labor is required
- Time consuming for complex geometries
- Smoke and fire hazards
- Anisotropic properties along planar and building direction
- Layer height is only dependent on thickness of sheets
- Wooden parts having thin cross-sections have poor strength and also absorb moisture
- Breaking out of parts may be difficult
- Material wastage may be high if final product does not make use of full build volume
- For internal geometries, extra material removal may be quite difficult.

11.3.6 Design and Quality Aspects

LOM has the following design and quality aspects [6]:

- Layer thickness generally varies between 0.05 and 0.2 millimetres
- Least sectional dimensions obtained is 0.2 millimetres
- Restricted suitability for components which can be subjected to shear forces along build plane axis owing to poor bonding of sheets
- As process generates smoke, the build chamber must be sealed
- Achievable tolerances are ±0.1 to ±0.25 millimetres
- Roughness of surfaces are typically in order of 30–40 µm Ra
- May require additional finishing processes to enhance surface finish.

11.3.7 Applications of LOM

The common applications of LOM processes include:

- Prototypes not to be utilized for functional applications
- Patterns and cores of castings
- Tool models
- Decorative products.

In addition to these popular applications, LOM is used for making several prototypes/products. A few examples of the same are illustrated in Figures 11.2 and 11.3.

FIGURE 11.2
LOM Processed Components Made of Pre-Ceramic Papers Filled with Ceramic Powder: (a) Turbine Rotor, 60 mm Diameter Made of Al$_2$O$_3$; (b) SiC Gear Wheel, 50 mm diameter. (*Source:* [7]. Travitzky et al., Preceramic paper-derived ceramics. *Journal of the American Ceramic Society*, 2008, 91(11): pp. 3477–3492, reproduced with permission from Wiley.)

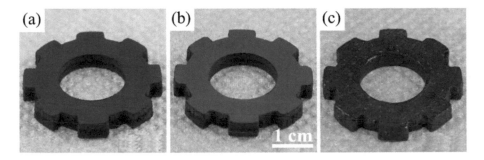

FIGURE 11.3
LOM Processed 3D Gear with MAX Phase Ti$_3$SiC$_2$: (a) In Green State; (b) After Sintering in Ar; (c) After Silicon Infiltration. (*Source:* [8]. Reprinted from Krinitcyn et al., Laminated object manufacturing of in-situ synthesized MAX-phase composites. *Ceramics International*, 43(12): pp. 9241–9245, Copyright 2017, with permission from Elsevier.)

11.4 Ultrasonic Consolidation

Ultrasonic consolidation (UC), also termed ultrasonic additive manufacturing (UAM), is another main sheet lamination process. It is a solid state AM technique which joins metallic materials (similar or dissimilar in the form of thin sheets or foils) in layer-by-layer fashion using the principles of ultrasonic welding (USW). It is a hybrid AM technique which combines the additive and subtractive manufacturing principles. In this technique, metallic foils are initially joined using USW and then selective machining is

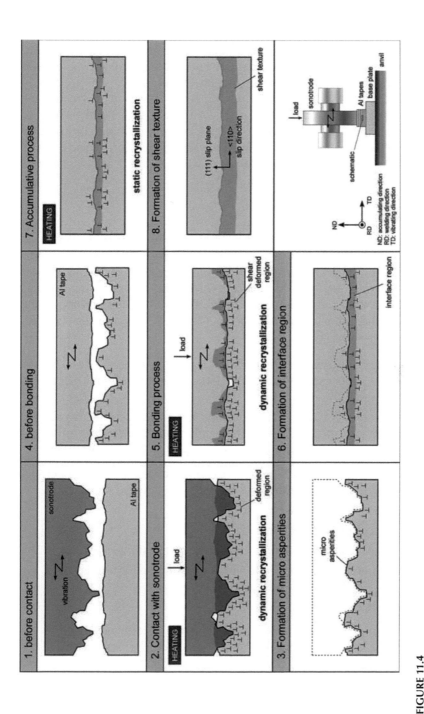

FIGURE 11.4
Stepwise Illustration of Microstructural Evolution during UAM. (*Source:* [10]. Reprinted from Fujii et al., Quantitative evaluation of bulk and interface microstructures in Al-3003 Alloy builds made by very high power ultrasonic additive manufacturing. *Metallurgical and Materials Transactions A*, 2011, 42(13): pp. 4045–4055, Springer, with permission from Springer Nature.

performed using CNC integrated with a UAM system to obtain near net shape of objects. It was invented and patented by White [9] and then commercialized by Solidica Inc. (USA) in 2000, a company founded by Dr. White. Now this technology is owned by Fabrisonic Inc. After its invention, several metals such as aluminium, copper and titanium alloys have been processed using UAM. A schematic illustration of the UAM process is shown in Figure 11.4, which presents different stages occurring during component fabrication via the UAM route. An interesting point to note is that if examined carefully the figure has three columns, where column one shows micro-asperity formation, column two shows bonding during UAM and column three shows the accumulative process. A detailed discussion of UAM is presented in Chapter 15 of this book.

11.4.1 Benefits of UAM

Being a solid state process, UAM has several benefits over other AM processes. Most of the metallic raw material-based AM systems, i.e. metal AM (MAM) modellers, basically work at a temperature near to or above the melting point temperature of materials, which leads to poor mechanical properties. UAM works at a temperature less than half of the melting point temperature of the material under process. Owing to this, it offers various unique benefits as summarized below:

- Due to lower temperature generation during the process, the fabricated components exhibit less distortion, embrittlement and dimensional changes.
- With the combination of additive and subtractive processes, it is capable of developing complicated multifunctional metallic 3D components.
- Suitable for fabricating multi-material structures.
- Ability to embed functional/sensitive components among the sheet/foil layers during the bonding of structure into a dense matrix.
- Ability to embed multiple types of fibers.
- Higher deposition/production rates.
- Ability to fabricate components with large dimensions.
- No atmospheric control is needed to take care of oxidation issues.

11.4.2 Drawbacks of UAM

Despite the several benefits offered, UAM has not established itself as an attractive alternative owing to the major issue of the "height-to-width ratio." As per Robinson et al. [11], if the height of the build approaches the width of the structure then failure of bonding occurs between the foil and base plate and subsequent layers cannot be added. However, it is independent of length of the build.

11.4.3 Applications of UAM

Despite the several benefits of UAM, shifting of UAM from laboratories to industrial applications is very rare. Current UAM applications are mainly in rapid tooling. Being a low temperature process, UAM has the following current and probable applications:

- Embedding of electronic structures into metal matrices
- Embedding of fibers
- Dissimilar metallic laminates and functionally gradient materials for specific application, and so on.

11.5 Summary

As has been evidenced through a detailed discussion on sheet lamination-based AM techniques, a wide spectrum of these is commercially available today. However, the two main classes are LOM and UAM. UAM is also covered in detail in Chapter 15 on hybrid additive manufacturing techniques. One remarkable and noteworthy trait of these processes is that they possess speeds of layer-wise processes but utilize only point-wise energy source. This can mainly be attributed to the fact that the sheet utilized as raw material is trimmed only at the outlines and does not require its complete melting. Post-processing requirement in such processes is also quite limited. Enhanced speeds, better materials, better bonding as well as support strategies, and improved placement and trimming techniques are some areas which are being continuously researched to arrive at improved variants of sheet lamination-based AM processes. Much development is still required in this field before they can attain the central capacity as the leading and most successful AM techniques.

In this chapter, an attempt has been made to cover all the necessary aspects of sheet lamination-based AM techniques. The next chapter covers the directed energy deposition-based AM processes.

References

1. Silbernagel, C., *Additive Manufacturing 101–6: What is sheet lamination?* Available from http://canadamakes.ca/what-is-sheet-lamination?, accessed January 29, 2109.
2. Kunieda, M., Manufacturing of laminated deep drawing dies by laser beam cutting. Proceedings of ICTP, 1984, 1: p. 520.

3. Dolenc, A., *An overview of rapid prototyping technologies in manufacturing*, 1994, Citeseer.
4. Upcraft, S., Fletcher, R., The rapid prototyping technologies. *Assembly Automation*, 2003, **23**(4): pp. 318–330.
5. Chen, Z., Li, Z., Li, J., Liu, C., Lao, C., Fu, Y., Liu, C., Li, Y., Wang, P., He, Y., 3D printing of ceramics: a review. *Journal of the European Ceramic Society*, 2019, **39**(4): pp. 661–687.
6. Swift, K. G., Booker, J. D., Rapid prototyping processes, in *Manufacturing process selection handbook*, K. G. Swift and J. D. Booker, Editors, 2013, Butterworth-Heinemann, pp. 227–241.
7. Travitzky, N., Windsheimer, H., Fey, T., Greil, P., Preceramic paper-derived ceramics. *Journal of the American Ceramic Society*, 2008, **91**(11): pp. 3477–3492.
8. Krinitcyn, M., Fu, Z., Harris, J., Kostikov, K., Pribytkov, G. A., Greil, P., Travitzky, N., Laminated object manufacturing of in-situ synthesized MAX-phase composites. *Ceramics International*, 2017, **43**(12): pp. 9241–9245.
9. White, D., *Ultrasonic object consolidation*, U. Grant, Editor, 2003, Solidica.
10. Fujii, H. T., Sriraman, M. R., Babu, S. S., Quantitative evaluation of bulk and interface microstructures in Al-3003 Alloy builds made by very high power ultrasonic additive manufacturing. *Metallurgical and Materials Transactions A*, 2011, **42**(13): pp. 4045–4055.
11. Robinson, C. J., Zhang, C., Janaki Ram, G. D., Siggard, E. J., Stucke, B., Li, L., Maximum height to width ratio of freestanding structures built using ultra-sonic consolidation, in *Proceedings of 17th solid freeform fabrication symposium*, 2006, Austin.

12

Additive Manufacturing Processes Utilizing Directed Energy Deposition Processes

12.1 Introduction

This is the seventh chapter of Section B of this book, which presents process specific details of various AM processes. In the previous chapter, readers were introduced to all the basic details of the AM processes utilizing sheet lamination techniques. This chapter deals with AM processes utilizing directed energy deposition (DED) processes and covers the general DED process description; laser-based DED techniques including direct laser deposition, laser-based DED techniques for 2D geometries and laser-based DED techniques for 3D geometries; applications of laser-based DED (LB-DED) including laser-assisted repair, laser cladding and electron beam-based DED processes; and advantages, limitations and applications of the DED process. This chapter concludes the discussion with a summary.

DED processes are a group of techniques which develop parts by melting the material when it is being deposited. According to the ISO/ASTM definition,

> DED is an additive manufacturing process in which focused thermal energy is used to fuse materials by melting as they are being deposited. "Focused thermal energy" means that an energy source (e.g., laser, electron beam or plasma arc) is focused to melt the materials being deposited.

Powder/wire is simultaneously fed along with focussed energy and thus, principally, this process contrasts with PBF where selective melting of the powder bed takes place.

12.2 Variants of Directed Energy Deposition

Before proceeding further, a quick overview of the main historical development of these processes is presented.

During 1994–1997, Sandia National Laboratories (New Mexico, USA) innovated a novel AM technique which they termed LENS (laser engineered net shaping). This was appreciably different from other existing AM techniques of the early 2000s and led to a spawning of multiple similar processes. One such process was DMD (2002, USA-based POM group). Different trade names are used for these techniques, like DLD, LC, LMD, DLF and so on, but each one of these is more or less a variation of the LENS process. However, one of these was LCVD, which came into being even before LENS (around the 1980s) but was not effectively utilized in developing parts till at least a decade after its inception. 3D welding (1990s, Germany) enabled fabricating parts using welders but its use was more or less limited till the 1990s. EBF3 utilizing electron beam and solid wire-based feedstock material was developed (NASA, 2002) which enabled part creation in space without the effect of gravity. A lot of research has been in process from 2010 onwards. A noteworthy point is that while the DED processes appear dissimilar to each other, the fundamentals are more or less similar for each of them.

There are many variants of DED technology. Several organizations have designed and developed different DED machines with varying heat sources and powder feeders. The most common are summarized in Table 12.1.

TABLE 12.1

Various Commercial DED Processes

Technique	Acronym	Source
Laser cladding	LC	[3]
Direct laser deposition	DLD	[4]
Direct laser fabrication	–	[4]
Direct metal deposition	DMD	[5]
Directed light fabrication	DLF	[6]
Electron beam additive manufacturing	EBAM™	Sciaky, Inc.
Laser forming	Lasform	[7]
Laser engineered net shaping	LENS	[8]
Laser powder fusion	LPF	[9]
Laser-aided direct-metal deposition	LADMD	[10]
Laser aided manufacturing process	LAMP	[11]
Laser chemical vapor deposition	LCVD	[12]
Laser consolidation	LC	[13]
Shape deposition manufacturing	SDM	[14]
Wire arc additive manufacturing	WAAM	[15]

Sources: [1, 2]. Saboori et al., An overview of additive manufacturing of titanium components by directed energy deposition: microstructure and mechanical properties. *Applied Sciences,* 2017, 7(9): p. 883; and Silbernagel, *Additive Manufacturing 101–4: What is directed energy deposition?* Available from: http://canadamakes.ca/what-is-directed-energy-deposition, accessed February 13, 2019.

12.3 Process Description

Material as well as an energy source are simultaneously introduced in DED processes. Direct melting of the material in the form of powder or filament occurs by means of the energy source at the deposition location. Creation of a melt pool occurs as the energy source interacts with the material. This pool undergoes instantaneous solidification with the retraction of the energy source from the deposition location. For continuous AM process, scaling up of this one-point depositing strategy takes place and the material is continuously fed. For creation of 2D patterns, precise control of X and Y axes also needs to be accomplished. The pattern of movement of X and Y axes governs the material deposition. For example, if they are to follow a specific raster pattern then the material will be deposited as per the same pattern and thus creation of a layer will take place. Now, after the complete creation of the layer, movement of the Z axis in the downward or upward direction will create the shifting of the build platform in a manner to enable deposition of the next layer upon the previous one. The process can be repeated for realizing the final product. This principle forms the basis of the entire class of DED processes.

An illustration showing the application of energy source/beam directly over the materials during deposition is shown in Figure 12.1.

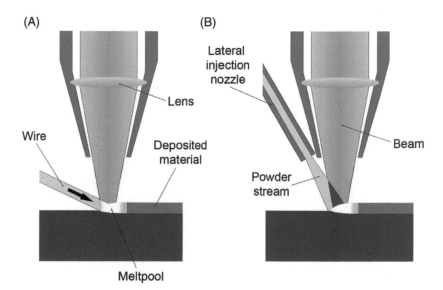

FIGURE 12.1
Wire and Powder Stream Directed Energy Deposition Process, (a) and (b) Respectively. (*Source:* [16]. Reprinted from Molitch-Hou, Overview of additive manufacturing process, in *Additive Manufacturing*, J. Zhang and Y.-G. Jung, Editors, pp. 1–38, Copyright 2018, with permission from Elsevier.)

12.4 Classification of DED Techniques

DED techniques can be classified according to the source of energy. Based upon this criterion, DED can be laser-based, electron beam-based, and plasma arc-based. However, owing to the much less popularity of plasma arc-based DED techniques, the other two categories of DED are discussed below.

12.4.1 Laser-Based DED Techniques

There are several laser-based DED techniques, hereafter LB-DED. Most important amongst them is direct laser deposition (DLD) which can be sub-classified as laser-based DED techniques for 2D geometries and laser-based DED techniques for 3D geometries. These are thoroughly discussed in the subsequent sections.

12.4.1.1 Direct Laser Deposition

Direct laser deposition (DLD), as a DED process, is a special class of the laser-based additive manufacturing process. A laser beam continuously irradiates the metal wire or powder preforms for their localized deposition. The substrate shifts suitably in the Z direction to enable creation of the entire 3D part from zero medium. In these systems, a focussed laser beam is oriented along with a deposition head that may consist of either one or many nozzles. As soon as the deposition of the model material particles takes place along the deposition profile, an ample quantum of thermal energy is supplied by the laser to accomplish their melting leading to the creation of a melt pool. A heat-affected zone of variable penetration depth thus develops. Movement of the build plate with the help of CNC with respect to the deposition head takes place on a complete deposition of the first layer to accommodate deposition of the subsequent layers and thus creation of the complete part. The process can be thermally monitored via different devices like infrared cameras and pyrometers and the data obtained can be further utilized for feedback or collection. Fixed build plates with material composition identical to that of the preform are typically used in DLD systems. Substrate can be thought of as the stage for the process. Shearing off the parts from the substrate is required for their removal. Parts using a wide spectrum of metallic and ceramic materials like Inconel 625, stainless steels, H-13 tool steel, Ti alloy, Cr, etc. can be successfully developed via this route. WC-Co cernet can be fabricated using this DED process.

12.4.1.1.1 Process Parameters of DLD

There are a number of important DLD process parameters. The nature of these is material and machine specific. They are also dependent upon the ambient and operating conditions. A few of these are: laser/substrate relative velocity (traverse speed), which is an indicator of time required to complete the part; laser scanning pattern, which should be adjusted prior to the process; laser power, which ranges from 100 to 5000 W for a beam diameter of 1 mm; laser beam diameter; hatch spacing; particle/powder feed rate; and interlayer idle time. These are adjusted for controlling structural traits, graded properties or porosities.

Many LB-DLD techniques have come up over the last two decades and vary in success and consumer adoption. A list of these has already been presented in Table 12.1.

Powder-based LB-DLD comprises a chain of occurrences that are related to each other and which occur in an extremely short time-scale. Several feasible instantaneous trajectories for momentum as well as energy are available, the occurrence of which can be either subsequent or in-parallel, as presented by Figure 12.2. A few of the aforementioned physical events, as per their tentative occurrence are: laser delivery; material delivery (energy/dynamics); laser–powder–gas interplay; formation of melt pool and its energy; stability as well as morphology; heat losses to environment through radiation/convection; solidification; conduction inside part; thermal cycling and part-substrate conduction. "Sub-events" details of each category are provided in Figure 12.2 and the same have been thoroughly researched either as LB-DED alone or as a part of conventional fabrication techniques like laser welding/cladding. Figure 12.2 highlights the major findings related to the sub-events.

DLD techniques can be sub-classified as those used for: (1) creation of 2D geometry like modifying/coating surfaces, cladding, etc. and (2) creation of 3D parts.

12.4.1.1.2 Laser-Based DED Techniques for 2D Geometries

Laser melt injection and laser cladding are the two LB-DED techniques used for generating 2D parts. A laser is utilized to melt the model material (powder/filament form) in both these processes. Both processes use a laser for melting the incoming material, which is either a powder or a filament. The process is accomplished in two steps. The first step includes preplacement of a narrow powder layer and its subsequent compaction over the surface. The next step involves scanning of the preplaced powder with high-powered laser light causing it to melt. With the development of computerized motion control systems, these two steps have been coupled into a one-step process where injection of powder upon the surface and its exposure to laser radiation occurs at a common location instantaneously, thereby causing its melting and solidification.

12.4.1.1.3 Laser-Based DED Techniques for 3D Geometries

These systems have CNC operated X and Y for carrying out raster scanning and a set-up for delivering laser and model material. To enable deposition of raster scans over previous layers, a laser beam is mounted along the

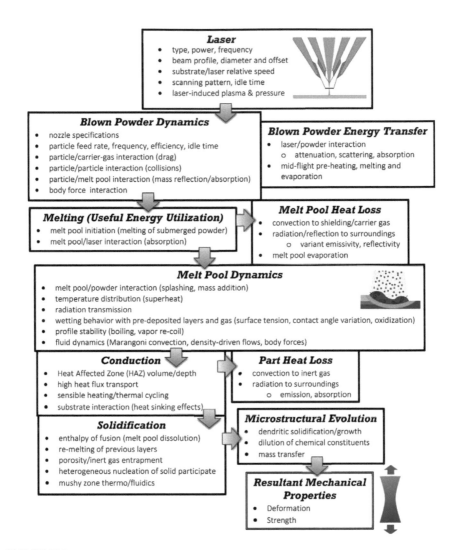

FIGURE 12.2
Various Events during DLD Process. (*Source:* [17]. Reprinted from Thompson et al., An overview of direct laser deposition for additive manufacturing. Part I: transport phenomena, modeling and diagnostics. *Additive Manufacturing*, 8: pp. 36–62, Copyright 2015, with permission from Elsevier.)

Z axis. A variety of companies provide systems based on these processes which can process different metallic as well as ceramic raw materials. Common techniques under this class include Laser Engineered Net Shaping (LENS), that was initially introduced at Sandia National Laboratories. LENS has quickly become a key powder-based LB-DED and is being quickly adopted by research and industry. Its operating principle is shown in Figure 12.3. Owing to LENS being a pioneer LB-DED, the term is sometimes used synonymously for it. An actual image of an Optomec LENS® 860 system is presented in Figure 12.4. Many net shaped components can be obtained using LENS technology. One such example (a functional hip stem with tailored porosity level) is presented in Figure 12.5.

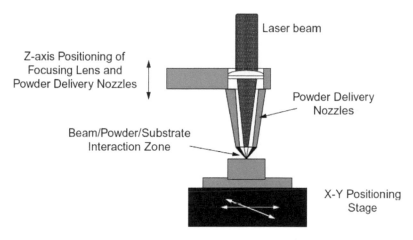

FIGURE 12.3
Schematic Representation of LENS™ Process. (*Source:* [18]. Reprinted from Krishna et al., Low stiffness porous Ti structures for load-bearing implants. *Acta Biomaterialia,* 3(6): pp. 997–1006, Copyright 2007, with permission from Elsevier.)

FIGURE 12.4
Optomec LENS® 860 System. (*Source:* Courtesy of Optomec Inc.)

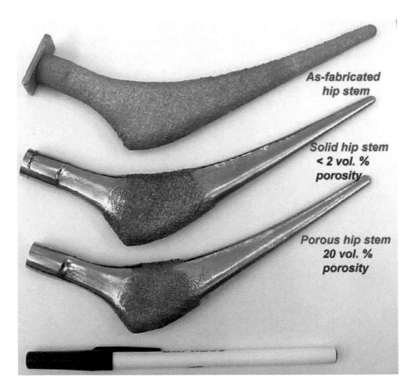

FIGURE 12.5
Net Shape, Functional Hip Stems with Designed Porosity Fabricated Using LENS™. (*Source:* [18]. Reproduced with permission from Krishna et al., Low stiffness porous Ti structures for load-bearing implants. *Acta Biomaterialia*, 2007, 3(6): pp. 997–1006.)

12.4.1.1.4 Applications of Laser Based DED (LB-DED)

LB-DED is successful in directly building near-net shaped 3D parts that can be utilized in a wide range of industrial parts. LB-DED also enables addition of the adequate amount of material at prescribed locations and has several other special capacities that include repairing, remanufacturing, cladding, depositing designated material at specified locations, etc.

- *Laser-assisted repair:* As compared to purchasing new parts, the cost incurred in repair of worn components is always less. Repairing damage as well as adding valuable details for enhancing features are some unique applications of laser-based DED processes. One such example is presented in Figure 12.6. No special changes in the AM set-up are required for repairing or adding features. However, exact and accurate knowledge of the damage to be repaired is mandatory. Normally the material required for the repair is same as the base material, but owing to various conventional as well as AM techniques applied for repairing, there is significant microstructural variation in the base material and repair deposit. Thus, employment of heat treatment to homogenize microstructures is required. Sometimes, this phenomenon of variation of component material properties from deposited material is also observed when functional areas or features are added. A concrete knowledge and hence analysis of metallurgical phenomenon of materials is mandatory to understand these processes. While it is impossible to combine some materials, others may require special techniques of deposition. Also, some materials can be fully deposited while others may only be deposited partially. Critical repair of several expensive parts has been accomplished using LENS. Performance of existing parts can be enhanced using simple laser cladding techniques.

- *Laser cladding:* Deposition of newer layers on substrate is usually carried out for repairing build-up in case of both cladding and hard facing. More than one-layer deposition can be done for achieving intricate geometry shapes. Cladding and hard facing are two variants of LMD/DMD and are utilized to modify material surface properties as also for repairing/manufacturing multi-layer coatings. Carbon dioxide lasers have been found most successful for both these variants. If the flexible LMD/DMD system principles are combined with those of fiber lasers, then a lot of overall innovative improvement can be achieved.

FIGURE 12.6
Example of Part Repairing Using LENS. (*Source:* Courtesy of Optomec Inc.)

Robust big DMD workstations (POM Group Inc., DMD 105D) have been developed to hard face, repair/clad large dies, molds as well as other parts. Fiber lasers with short wavelength achieve equivalent deposition rates while consuming 50% wattage of a CO_2 laser. Thus, the same production rate is favorably achieved with considerably less stress developed in the cladded part. Surfaces can be left after deposition or can be ground to obtain the finishing dimension.

Ability to precisely deposit multiple materials at different locations within the same part is a distinguishing characteristic of DED processes. This facilitates the utilization of DED as a means for development of artificial materials, i.e. an exquisite category of materials whose existence is not naturally possible and which are designed for optimal performance. Thus, ability to choose performance can be imparted to the material system with the help of DED processes.

12.4.2 Electron Beam-Based DED Processes

Electron beam-based DED processes (EB-DED, or EBF3 processes) were developed at NASA, Langley, USA. They are mainly used for fabrication as well as repair of both terrestrial and future space-oriented aerospace parts. These systems utilize an electron beam for thermal energy and a wire feeder. These possess the capability to output high (less accurate) and low (more accurate) deposition rates at high and low current rates respectively. Electron beams can effectively convert electrical energy as compared to lasers and are therefore favored for space-oriented applications. However, electrical resources are conserved in the case of lasers. Also, the EB-DED process requires a vacuum, as compared to the inert atmospheres required in laser-based processes, which again makes their candidature strong for space applications. Wire-based processes are preferred over powder ones owing to the difficulty in powder handling in space.

Some EB-DED (Sciaky, USA) use wire feedstock and are built inside extremely big vacuum chambers. They allow depositions of more than 6 m in build volumes. They possess distinct capabilities to quickly build big, heavy deposits and thus large rib-on-plate type, etc. deposits can be obtained (used in the aerospace sector). These can eliminate the appreciably high lead-times required for forging components when they are used to substitute them.

12.5 Advantages of DED

Several techniques come under category of DED-AM technology, that is why benefits are largely dependent on the type of process. Some common advantages of DED are:

- Ability to repair and clad damaged parts
- Relatively higher deposition rates as compared to PBF processes
- Ability to process large build volumes
- Ability to produce graded structures consisting of different materials
- Ability to repair and remanufacture.

12.6 Drawbacks of DED

Despite various advantages, DED technology has some inherent drawbacks, including:

- Need for support structure
- Accuracy and surface finish are of relatively less quality as compared to contemporary AM processes
- Owing to merging of energy exposure and melting cycles, modelling of such systems becomes rather complicated.

12.7 Summary

For electron beam DED, a vacuum is required, in contrast to laser-based DED techniques, but they possess the advantage of better energy density as well as efficiency compared to DLD. Electron beam-based techniques are a more lucrative means to build/repair parts that exhibit high oxygen reactivity and also for fabricating functional parts. Plasma deposition (PDM)-based fabrication is also showing a lot of promise.

In this chapter, an attempt has been made to cover all the necessary aspects of directed energy deposition-based AM techniques. This is the last chapter of Section B. Section C covers design and quality aspects of AM processes and consists of Chapters 13 and 14.

References

1. Saboori, A., Gallo, D., Biamino, S., Fino, P., Lombardi, M., An overview of additive manufacturing of titanium components by directed energy deposition: microstructure and mechanical properties. *Applied Sciences*, 2017, 7(9): p. 883.
2. Silbernagel, C., *Additive Manufacturing 101–4: What is directed energy deposition?* Available from: http://canadamakes.ca/what-is-directed-energy-deposition, accessed February 13, 2019.

3. Weerasinghe, V. M., Steen, W. M., Laser cladding by powder injection, in *Proceedings of the 1st International Conference on Lasers in Manufacturing*, 1983.

4. Gu, D. D., Meiners, W., Wissenbach, K., Poprawe, R., Laser additive manufacturing of metallic components: materials, processes and mechanisms. *International Materials Reviews*, 2012, **57**(3): p. 133–164.

5. Mazumder, J., Choi, J., Nagarathnam, K., Koch, J., Hetzner, D., The direct metal deposition of H13 tool steel for 3-D components. *JOM*, 1997, **49**(5): pp. 55–60.

6. Milewski, J., Lewis, G. K., Thoma, D. J., Keel, G. I., Nemec, R. B., Reinert, R. A., Directed light fabrication of a solid metal hemisphere using 5-axis powder deposition. *Journal of Materials Processing Technology*, 1998, **75**(1–3): pp. 165–172.

7. Arcella, F. G., Froes, F. H., Producing titanium aerospace components from powder using laser forming. *JOM*, 2000, **52**(5): pp. 28–30.

8. Keicher, D. M., Miller, W. D., LENSTM moves beyond RP to direct fabrication. *Metal Powder Report*, 1998, **12**(53): pp. 26–28.

9. Shamsaei, N., Yadollahi, A., Bian, L., Thompson, S. M., An overview of Direct Laser Deposition for additive manufacturing; Part II: Mechanical behavior, process parameter optimization and control. *Additive Manufacturing*, 2015, **8**: pp. 12–35.

10. Choi, J., Process and properties control in laser aided direct metal/materials deposition process, in *ASME 2002 International Mechanical Engineering Congress and Exposition*, 2002, American Society of Mechanical Engineers.

11. Zhang, J., Liou, F., Adaptive slicing for a multi-axis laser aided manufacturing process. *Journal of Mechanical Design*, 2004, **126**(2): pp. 254–261.

12. Williams, K., Maxwell, J., Larsson, K., Boman, M., Freeform fabrication of functional microsolenoids, electromagnets and helical springs using high-pressure laser chemical vapor deposition, in *Technical Digest. IEEE International MEMS 99 Conference. Twelfth IEEE International Conference on Micro Electro Mechanical Systems (Cat. No. 99CH36291). 1999.

13. Xue, L., Ul Islam, M., *Laser Consolidation: A Novel One-Step Manufacturing Process for Making Net-Shape Functional Components*, 2019.

14. Fessler, J., Merz, R., Nickel, A. H., Prinz, Fritz B., Weiss, L. E., Laser deposition of metals for shape deposition manufacturing, in *Proceedings of the solid freeform fabrication symposium*, 1996, Citeseer.

15. Derekar, K. S., A review of wire arc additive manufacturing and advances in wire arc additive manufacturing of aluminium. *Materials Science and Technology*, 2018, **34**(8): pp. 895–916.

16. Molitch-Hou, M., Overview of additive manufacturing process, in *Additive manufacturing*, J. Zhang, Y-G. Jung, Editors, 2018, Butterworth-Heinemann, pp. 1–38.

17. Thompson, S. M., Bian, L., Shamsaei, N., Yadollahi, A., An overview of direct laser deposition for additive manufacturing. Part I: transport phenomena, modeling and diagnostics. *Additive Manufacturing*, 2015, **8**: pp. 36–62.

18. Krishna, B. V., Bose, S., Bandyopadhyay, A., Low stiffness porous Ti structures for load-bearing implants. *Acta Biomaterialia*, 2007, **3**(6): pp. 997–1006.

Section C

Material, Design and Related Aspects of Additive Manufacturing Processes

13

Materials for Additive Manufacturing

13.1 Introduction

Additive manufacturing came into existence around the 1980s and has since been continuously evolving till now. AM is based on fabricating layered arte-facts and has the unique potential of delivering highly customized parts. The need for supply chain management is totally eliminated, material wastage is tremendously reduced and manufacturing lead times are greatly shortened by the compression of design cycle, thereby making it a subject of great interest amongst manufacturers. Since AM is a direct output-oriented manufacturing strategy, much energy and fuel consumption saving can be obtained, therefore reducing carbon footprints and greenhouse gases. Though AM was initially seen as a strategy to complement conventional manufacturing, today it has bypassed it in many applications. To fully understand the AM process, there are various important aspects that need consideration. These include defini-tions, terminology, classification, historical development, various commercial AM techniques, process details, raw materials, applications, trends, variants, advancement and several other aspects.

This is the first chapter of Section C, which covers design and quality aspects of AM processes. This chapter presents a detailed discussion of the various materials in additive manufacturing; their forms/state including poly-mers, metals, ferrous alloys, Ti-alloys, ceramic materials, composite materials, etc.; material binding mechanisms in AM including binding using secondary phase assistance, binding using chemical induction, binding using solid state sintering (SSS) and binding using liquid fusion; and defects in AM parts including balling phenomena, porosity defects, cracks, distortion, inferior surface finish, etc. This chapter concludes the discussion with a summary.

13.2 Materials for AM

Materials play a major role in almost all of the AM techniques. A variety of materials is utilized for AM processes and research is in progress in developing newer materials for AM applications. Initially, the AM technique was originated around polymers, paper laminates and waxes. Plastics were the first group of materials to be utilized for AM and these still constitute the major raw materials for AM processes.

Subsequently, with the advances in AM processes other materials like metals, composites, ceramics, etc. have been introduced. As a result, a broad material spectrum can currently be processed via AM techniques; these include polymers, metals, ceramics, functionally graded materials, smart materials, composite materials, hybrids, etc. A summary of AM materials is presented in Figure 13.1.

To date there are more than 100 AM techniques [1]. These techniques are different in the ways in which layers are deposited, their working principles and the materials that can be processed via these techniques. AM processes can be classified in several ways as stated by Rathee et al. [1]. However, the classification discussed under Section 3.2 and detailed in Table 3.7 as well as Figure 3.1 of Chapter 3 of this book is predominantly preferred over other classifications. Based upon this classification as per ASTM guidelines, there are seven classes of AM processes: BJ, DED, ME, MJ, PBF, SL and VP processes.

Basic principles, examples of technology, advantages, drawbacks, materials utilized and tool manufacturing companies of these seven AM categories are summarized in Table 13.1. A detailed discussion of these individual techniques has been presented in Section B of this book. As listed in Table 13.1, the types of materials suitable for particular AM

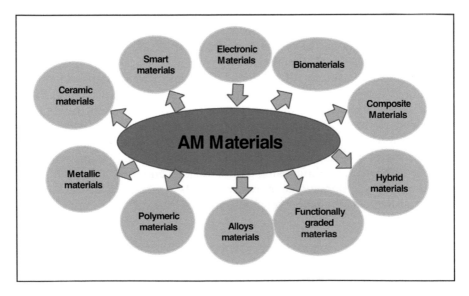

FIGURE 13.1
Classification of AM Materials.

techniques are different. For example, powder bed fusion processes can suitably process all types of materials such as polymers, metals, hybrids, ceramics and composites, while vat photo polymerization processes can suitably process only polymers and ceramics.

13.2.1 Forms/State of Materials for AM Processes

In addition to the suitability of types of materials for AM processes, their state is also important. For AM processes, the feedstock should have the form (e.g. liquid, powder, wire, sheet) which is compatible with the particular process. Some techniques, such as stereolithography, cure liquid materials, some techniques, such as ultrasonic AM, join metallic solid sheets to fabricate 3D parts, while other techniques, such as FDM, SLS, melt the materials to develop the layers. In general, the fluid state of material produces better deposition in AM techniques. In this way, polymer and polymer-based products such as polymer-based composites (PMCs), functionally graded materials, hybrids, etc. offer convenience owing to their lower processing/melting temperatures during AM processing. Metals and ceramics have higher melting point temperatures as compared to polymers and the ease of processing is basically in order of polymers > metals > ceramics. The bonding in ceramics as well as in metals is not as easy as in the case of polymers owing to their high melting point temperatures. Thus, each AM technique has its own pros and cons and offers suitability to different forms of materials. On the basis of the form of materials suited for particular techniques, AM processes can be categorized as presented in Figure 13.2. In addition, the compatibility of commonly used

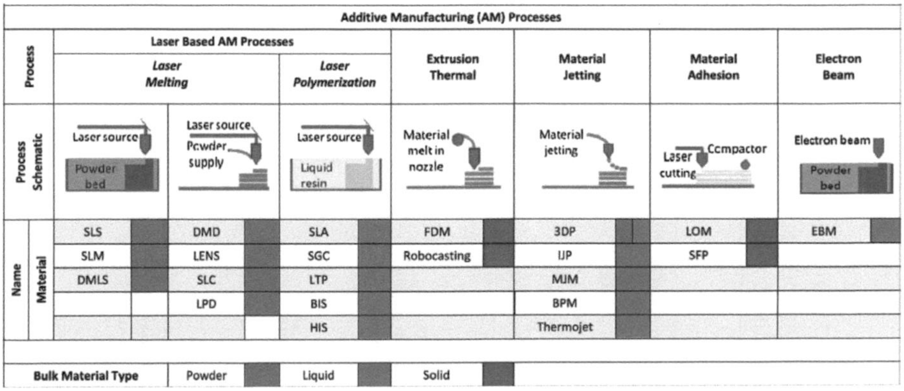

FIGURE 13.2
AM Categorization on Basis of Form of Materials. (*Source:* [3]. Bikas et al., Additive manufacturing methods and modelling approaches: a critical review. *The International Journal of Advanced Manufacturing Technology*, 2016, 83(1): pp. 389–405, under Creative Commons Licence.)

TABLE 13.1

Basic Principles, Materials, Advantages, Disadvantages, Typical Build Volumes and Tool Manufacturer of Seven ASTM Categories of AM: Binder Jetting (BJ); Directed energy deposition (DED); Material Extrusion (ME); (4) Material Jetting (MJ); Powder Bed Fusion (PBF); Sheet Lamination (SL); and Vat Photopolymerization (VP)

ASTM category	Basic principle	Example technology	Advantages	Disadvantages	Materials	Build volume (mm × mm × mm)	Tool manufacturer/country
BJ	Liquid binder/s jet printed onto thin layers of powder. The part is built up layer by layer by glueing the particles together	• 3D inkjet technology	• Free of support/substrate • Design freedom • Large build volume • High print speed • Relatively low cost	• Fragile parts with limited mechanical properties • May require post-processing	• Polymers • Ceramics • Composites • Metals • Hybrid	Versatile (small to large) X = <4000 Y = <2000 Z = <1000	ExOne, USA PolyPico, Ireland
DED	Focused thermal energy melts materials *during* deposition	• Laser deposition (LD) • Laser Engineered NetShaping (LENS) • Electron beam • Plasma arc melting	• High degree control of grain structure • High-quality parts • Excellent for repair applications	• Surface quality and speed requires a balance • Limited to metals/metal-based hybrids	• Metals • Hybrid	Versatile X = 600–3000 Y = 500–3500 Z = 350–5000	Optomec, USA InssTek, USA Sciaky, USA Irepa Laser, France Trumpf, Germany
ME	Material is selectively pushed out through a nozzle or orifice	• Fused deposition modelling (FDM)/fused filament fabrication (FFF), fused layer modelling (FLM)	• Widespread use • Inexpensive • Scalable • Can build fully functional parts	• Vertical anisotropy • Step-structured surface • Not amenable to fine details	• Polymers • Composites	Small to medium X = <900 Y = <600 Z = <900	Stratasys, USA
MJ	Droplets of build materials are deposited	• 3D inkjet technology • Direct ink writing	• High accuracy of droplet deposition • Low waste • Multiple material parts • Multicolor	• Support material is often required • Mainly photopolymers and thermoset resins can be used	• Polymers • Ceramics • Composites • Hybrid • Biologicals	Small X = <300 Y = <200 Z = <200	Stratasys, USA 3D Systems, USA PolyPico, Ireland 3Dinks, USA WASP, Italy

	Description	Technologies	Advantages	Challenges	Material types	Build volume	Manufacturers
PBF	Thermal energy fuses small region of powder bed of build material	• Electron beam melting (EBM) • Direct metal laser sintering (DMLS) • Selective laser sintering/melting (SLS/SLM)	• Relatively inexpensive • Small footprint • Powder bed acts as integrated support structure • Large range of material options	• Relatively slow • Lack of structural integrity • Size limitations • High power required • Finish depends on precursor powder size	• Metals • Ceramics • Polymers • Composites • Hybrid	Small X = 200–300 Y = 200–300 Z = 200–350	ARCAM, Sweden EOS, Germany Concept Laser Cusing, Germany MTT, Germany Phoenix System Group, France Renishaw, UK Realizer, Germany Matsuura, Japan, USA Voxeljet, USA 3Dsystems, USA
SL	Sheets/foils of materials are bonded	• Laminated object manufacturing (LOM) • Ultrasound consolidation/ultrasound additive manufacturing (UC/UAM)	• High speed • Low cost • Ease of material handling	• Strength and integrity of parts depend on adhesive used • Finishes may require post-processing • Limited material use	• Polymers • Metals • Ceramics • Hybrids	Small X = 150–250 Y = 200 Z = 100–150	3D systems, USA MCor, Ireland
VP	Liquid polymer in a vat is light-cured	• Stereo lithography (SLA) • Digital light processing (DLP)	• Large parts • Excellent accuracy • Excellent surface finish and details	• Limited to photopolymers only • Short shelf life, poor mechanical properties of photopolymers • Expensive precursors/slow build process	• Polymers • Ceramics	Medium X <2100 Y <700 Z <800	Lithoz, Austria 3D Ceram, France

Source: [2]. Tofail et al., Additive manufacturing: scientific and technological challenges, market uptake and opportunities. *Materials Today*, 2018, 21(1): pp. 22–37.

Notes: Build volumes are rounded to nearest number for convenience.
Material types have been ranked in order of suitability and common use.

individual AM techniques with the type and state of materials is shown in Figure 13.3. A detailed classification and corresponding AM processes have already been presented on the basis of the form/state of raw material in Section 3.2 of this book.

Subsequent sections present the discussion of materials and their issues for various AM processes.

13.2.2 Polymers

Polymers are widely utilized materials in AM techniques for manufacturing of plastics, polymer-based composites (PMCs) and polymer-based functionally graded materials (FGMs). This is mainly owing to the low melting or curing temperature, chemical stability and good flow ability (molten and softened state) of polymers. Polymers can be suitably processed via AM in the form of liquid, powders, sheets or filaments. Almost all fusion-based AM techniques can be applied to process polymers if adequate formulation

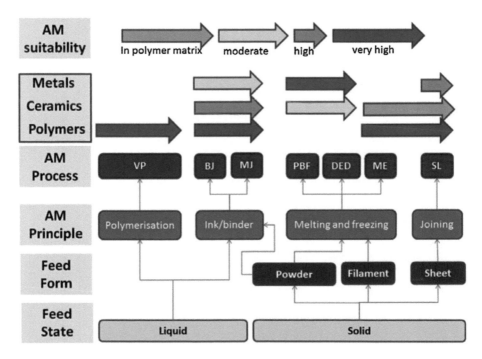

FIGURE 13.3
Suitability of AM Techniques (for Seven Categories of AM as per ASTM Classification) of Polymers, Metals and Ceramics in Different Feed Forms/States. (*Source:* [2]. Tofail et al., Additive manufacturing: scientific and technological challenges, market uptake and opportunities. *Materials Today*, 2018, 21(1): pp. 22–37, under Creative Commons Licence.)

of polymers is used. However, material jetting, extrusion and photopolymerization are the most common AM techniques to process polymers [4]. Thermoplastic polymers and UV-curable polymers are the most commonly processed AM materials.

The common thermoplastic polymers that are compatible with AM processes are acrylonitrile butadiene styrene (ABS), nylon, polylactic acid, polycarbonate and polyamide. Hardness at room temperatures is a characteristic feature of each. For 3D printing of polymers, all of the AM techniques utilize the same fundamental working principle of selective melting and solidification via constrained heating and deposition. PBF techniques are commonly utilized for UV-curable polymeric materials [5] where selective polymerization of polymeric monomer with a PI in vat is accomplished using a light source. In addition to thermoplastic and UV-curable polymers, soft polymers or elastomers can also be processed using AM techniques. Most commonly used elastomers are thermoset polymers. However, 3D printing of elastomers offers difficulty as compared to thermoplastic and UV-curable polymers and their 3D printing is mainly achieved using copolymer material systems, i.e. combination of elastomers and thermoplastics. In 3D printing of co-polymers, the thermoplastic component provides creep to thermoplastic elastomers for their easy processing. Figure 13.4 presents an overview of polymers, their fusing techniques and corresponding print technology used in AM.

13.2.3 Metals

MAM offers greater freedom for developing metallic components (simple as well as complex geometrical shaped) as compared to conventional manufacturing. Despite the fact that the research on MAM during the initial phase of invention of AM was quite limited, the last two decades witnessed an enormous growth in its domain. Nowadays, MAM has become the focus of research in order to meet the specific and functional requirements of components designed for industries such as aerospace, defence, automotive, etc. According to Wohler reports [7, 8], the number of sellers of commercial AM systems was reported as 97 in 2016, out of which almost 49% were involved with MAM.

Commercially used MAM techniques mainly include PBF and DED. These are chiefly based upon the principle of selective melting of metal powders to form the metal part. However, some newly developed MAM techniques, such as friction-based AM techniques [1, 9], binder jetting and cold spray-based AM techniques, are gaining attention in developing 3D components from a wide range of metallic materials. Metallic materials, such as (some grades of) ferrous alloys, nickel-based alloys, lightweight alloys (such as aluminium, magnesium, etc.), can be suitably processed via MAM techniques. A detailed discussion of these materials is presented below.

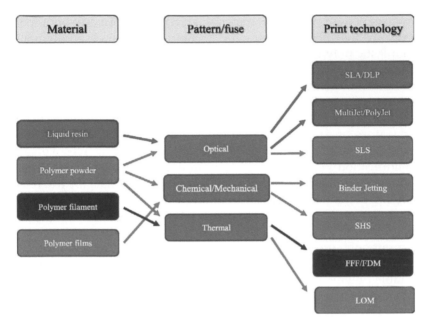

FIGURE 13.4
Overview of Polymers/Monomers Used in AM. (*Source:* [6]. Reprinted from Stansbury and Idacavage, 3D printing with polymers: challenges among expanding options and opportunities. *Dental Materials*, 32(1): pp. 54–64, Copyright 2016, with permission from Elsevier.)

- **Ferrous alloys:** The most commonly used ferrous alloys for AM are stainless steels of austenitic and precipitation hardening type. The austenitic stainless steels include 316-, 304-, 304L, 316L AISI type stainless steels [10]. These metal alloys are mainly deposited via PBF-laser and DED-laser AM processes.

- **Ti alloys:** Ti–6Al–4V is the most common individual alloy printed using MAM techniques. This is mainly due to the fact that this particular alloy has great applications in the bio-medical sector. In its purest form, Ti consists of two phases, namely the α and β phases. The α phase is stronger with lesser ductility, while the β phase is more ductile than α. $\alpha+\beta$ alloys exhibit higher strengths as well as formability. For achieving desired properties in titanium alloys, the α and β phases can be adjusted. For application as a bone mimics, the density of the part should be matching to the surrounding material which is generally high, thus requiring higher content of β phase. Al and V represent α and β stabilizing alloying elements inside the Ti–6Al–4V alloy. The formation of the less stable α phase occurs when the β phase is quenched. Thus, during 3D printing of

Ti–6Al–4V alloy, it is required to select appropriate printing conditions to obtain an adequate $\alpha+\beta$ phase combination for achieving desired properties in terms of strength, density, ductility and corrosion resistance.

13.2.4 Ceramic Materials

Several materials like ceramics or concrete generally do not find suitability in AM. This is mainly due to the fact that the individual particles of ceramics or concrete fail to fuse together after heating them to their melting points [11]. However, polymeric and metallic materials fuse together at their melting point temperatures. In comparison to polymers and metals, ceramic materials have extremely high melting point temperature which is a major challenge for most AM processes. AM techniques can produce components/parts from ceramics having comparable mechanical properties to parts produced using conventional manufacturing techniques. PBF-based AM techniques find suitability as cost effective methods for developing ceramic components. However, the unavailability of initial raw materials to obtain feedstock is a major limitation of AM of ceramics.

The applications of AM in advanced ceramics component fabrication to be utilized in tissue engineering and biomaterials are gaining popularity. The process of sintering of ceramics and their post-processing consumes additional energy and is considerably expensive. Still, the process of fabricating intricately shaped components and subsequent sintering has become attractive. Scaffolds made of ceramics manufactured via the AM route for bones and teeth in tissue engineering are gaining attraction owing to fast and convenient processing as compared to conventional methods such as casting and sintering.

Despite the existence of various AM techniques, SLS is the most common one for the processing of ceramic materials. However, the ceramic parts manufactured via SLS are more prone to crack formation owing to the involvement of thermal shocks of heating and cooling during the process. The layer-by-layer appearance of AM printed ceramic components/ structures is another major challenge, as depicted in Figure 13.5. In some applications such as tissue engineering, the layered appearance may not be important owing to the encapsulation of the final product, e.g. scaffolds. However, in some applications such as aerospace, buildings etc., flat surfaces are preferred.

13.2.5 Composite Materials

Composite materials are relatively new to AM processes as compared to polymers and metals. AM of metal matrix composites (MMCs) is gaining much attention owing to its ability to fabricate composites with enhanced properties as compared to their base metals. SLS and LMD are the main AM processes to develop composites such as WC-Co MMCs. Some problems which can exist in 3D printing of dense MMCs are due to entrapment of gases, the occurrence of gaps between matrix and reinforcement particles and so on.

The material compatibility of various raw materials with commonly used AM techniques is summarized in Table 13.2.

FIGURE 13.5
Layer-by-Layer Appearance of 3D Printed Concrete Structure. (*Source:* [12]. Reprinted from Gosselin et al., Large-scale 3D printing of ultra-high performance concrete: a new processing route for architects and builders. *Materials and Design*, 100: pp. 102–109, Copyright 2016, with permission from Elsevier.)

TABLE 13.2

Material Compatibility of Different AM Techniques

Popular AM techniques	Polymers	Metals	Ceramics	Composites
SLA	✓			✓
MJM	✓			✓
FDM	✓			
EBM		✓		
SLS	✓	✓	✓	✓
DMLS		✓		
LOM	✓	✓	✓	✓
UAM		✓		
LMD		✓		✓

Sources: [7]. Deloitte analysis; Wohlers Associates, *Additive manufacturing and 3D printing state of the industry*, 2014; Phil Reeves, 3D printing and additive manufacturing: extending your printing capability in true 3D. *Econolyst*, June 12, 2012; Justin Scott, IDA Science and Technology Policy Institute. *Additive manufacturing: status and opportunities*, March 2012. Graphic: Deloitte University Press | DUPress.com

13.3 Material Binding Mechanisms in AM

The efficacy of AM processes depends upon the efficiency and effectiveness with which the layers are bonded together. Binding techniques thus largely affect process speeds and fabricated part properties. The binding mechanisms in AM can be categorized into four categories as depicted in Figure 13.6. However, these categories are not exhaustive, but cover a clear trend. These mainly include:

- Binding using secondary phase assistance
- Binding using chemical induction
- Binding using solid state sintering
- Binding using liquid fusion.

13.3.1 Binding Using Secondary Phase Assistance

Many AM techniques utilize a secondary phase for mutual binding of material layers, for example, sheet lamination, binder jetting, SLS and so on. The secondary phase may be in the form of powder, coatings, liquids, etc. Addition of these secondary phases is accomplished by various means like nozzle injection, coatings and so on. Binding is done mainly by adhesives, liquid phase sintering, and evaporation and hydration.

Adhesive bindings are commonly used in binder jetting and sheet lamination techniques. In binder jetting, the adhesives (either in liquid or dry form) are added in the main material via automated nozzles or by

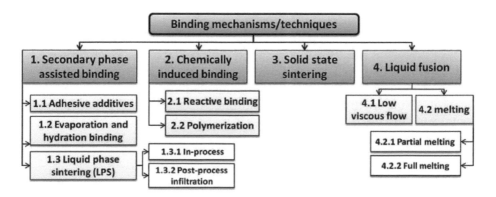

FIGURE 13.6
Binding Methods for AM Processes. (*Source:* [13]. Reprinted from Bourell et al., Materials for additive manufacturing. *CIRP Annals*, 66(2): pp. 659–681, Copyright 2017, with permission from Elsevier.)

embedding into the powder bed. Liquid binders contain binding material while dry binders react with the deposit after addition to the powder in the powder bed. In sheet lamination, coatings are generally applied and bonding of layers occurs due to application of heat and/or pressure.

13.3.2 Binding Using Chemical Induction

In chemical induction binding, no secondary phase constituents are used to bind the materials. AM processes such as material jetting, vat photo polymerization, etc. utilize the chemical reactivity of material for their binding.

13.3.3 Binding Using Solid State Sintering

Solid state sintering (SSS) is a thermal consolidation technique and is done at a temperature below the material's melting point temperature. It is utilized as a post-processing technique and is a diffusion binding process. During this, certain chemical and physical reactions take place. The atomic diffusion causes neck formation among adjacent particles that will grow with respect to time. This type of consolidation phenomena is mostly suited for the ceramic particles.

13.3.4 Binding Using Liquid Fusion

AM techniques mostly utilize the liquid fusion binding (LFB) mechanism, which includes low viscosity flows in polymeric and melting in metallic parts respectively. Fusion of polymer/plastic over previous layer takes place during low viscosity flow upon deposition. Some common examples can be wax droplets in the material jetting process, heated polymers deposited over the previous layer upon deposition in PBF.

Binding using melting is another consolidation mechanism which includes partial and full melting of metallic materials. During partial melting, one portion of material remains solid while the other portion is in melted form and the liquefied material spreads inbetween solid particles. Full melting is another prime binding mechanism which results in completely dense components, thus eliminating need for post-processing densification. Thus, melting is a main binding mechanism to produce dense metal parts in DED and PBF processes.

13.4 Defects in AM Parts

Despite the tremendous research progress in AM technology, some newly developed AM processes, especially MAM processes, are at their early stage and the relationship of basic processing parameters with microstructure and other properties is not fully understood. In the absence of optimized process parameters and material combinations, defects occur which result in poor mechanical properties of fabricated components. The subsequent sections present a brief overview of common defects for different materials and AM techniques.

13.4.1 Balling Phenomena

Balling is basically a phenomenon which occurs when the wetting of the underlying surface is not produced by the liquid material. This leads to a rough, bead-shaped scan track that reduces surface finish and increases pore formation. Such types of defect generally occur in laser sintering-based AM techniques. A typical example of balling phenomena is depicted in Figure 13.7.

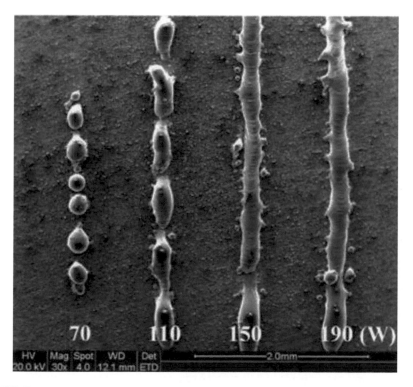

FIGURE 13.7
Illustration of Balling Phenomena during SLM Process. (*Source:* [14]. Reprinted from Li et al., Balling behavior of stainless steel and nickel powder during selective laser melting process. *The International Journal of Advanced Manufacturing Technology*, 2012, 59(9): pp. 1025–1035, Springer, with permission from Springer Nature.

13.4.2 Porosity Defects

Porosity and voids are commonly occurring flaws in AM parts, as the most utilized binding mechanisms are governed by changes in temperature under the action of capillary and gravity forces in absence of external force. The major reasons behind the porosity or voids defects are: (1) repeated formation of keyholes resulting in formation of voids; (2) entrapment of gases which results in microscopic pores in powder particles when they are atomized; (3) inadequate penetration of current layer of molten pool into previous layer or the substrate. Defects like keyholes induced porosity, absence of fusion pores and porosity due to gases are shown in Figure 13.8. Keyhole porosity is small in size and generally less than 100 μm.

13.4.3 Cracks

Crack defect is another main problem in AM parts which generally occurs in fusion-based AM techniques. During fusion-based AM techniques, metallic powders experience rapid melting and solidifications. Due to the rapid cooling rate, a high temperature gradient and subsequently large thermal residual stresses are generated in the fabricated part. The combination of large residual stresses and high temperature gradient result in initiation of cracks. Cracks are mostly prominent along grain boundaries.

This occurrence is most common along the boundary of grains and is chiefly due to temperature variations owing to different rates of contraction for each layer, i.e. substrate, solidifying and deposited layers. Liquation cracking is another defect and is mostly observed in a mushy/partially

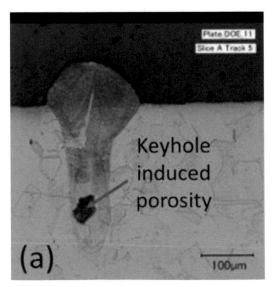

FIGURE 13.8
Porosity Defect (Keyhole Porosity) during MAM. (*Source:* [15]. Reprinted from King et al., Observation of keyhole-mode laser melting in laser powder-bed fusion additive manufacturing. *Journal of Materials Processing Technology*, 214(12): pp. 2915–2925, Copyright 2014, with permission from Elsevier.)

melted zone which experiences tensile force leading to liquid films acting as cracking points. Various other cracks can be similarly explained. These cracks are either appreciably long or relatively small. The process of delamination occurs when the layers separate mutually in the event of residual stress at the layer interface being more than the alloy's yield strength.

13.4.4 Distortion

Distortion is a kind of defect in AM parts which mainly originates due to the stresses developed in the material caused by the change in volume during shrinkage. Figure 13.9 shows some typical distortion defects.

13.4.5 Inferior Surface Finish

Inferior surfaces occur prominently in parts fabricated via the AM route. The most important single factor that can be held responsible for poor surface finish is the staircase effect. Other factors like part build orientation, roughly deposited beads (for instance, FDM coarse filament), reduced tool precision (for instance, considerably big heat-affected field in EBM), surface tension as well as balling (for instance, in PBFs), powder in semi-molten state (for instance, material clinging to down-facing surfaces during SLM), etc. are also responsible. High surface roughness is also due to aging of used material, e.g. extensively used polyamide powders in SLS result in poor surfaces resembling orange peel. Better surface finish can be imparted if small deposit beads (or powders) are utilized. This can also be achieved

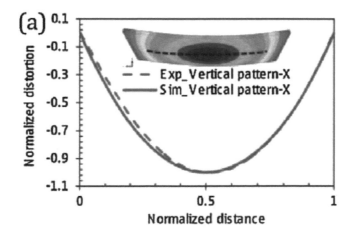

FIGURE 13.9
Typical Distortion Defect in AM Parts in SLM Processed Steel. (*Source:* [16]. Li et al., Fast Prediction and Validation of Part Distortion in Selective Laser Melting. *Procedia Manufacturing*, 2015, 1: pp. 355–365, under Creative Commons Licence.)

by decreasing layer thickness. These practices will however retard the production rates. The post-processing requirement is greatly increased in the case of inferior surface finish in intricate AM parts. Processes like grit blasting, mechanical grinding and so on for post-processing leads to increased cost of production.

13.5 Summary

AM raw materials can be of varied forms, including sheet, filament, paste, ink, gas, wire and powder. Materials like polymers, ceramics, composites, metals, alloys, functionally graded, smart, hybrid, etc. are widely utilized AM raw materials. However, anisotropy, mass customization, microstructural control, compositional control, variety, etc. aspects remain some major restraints upon the AM materials. Lack of regulatory issues is also a problem area that needs focus to broaden the spectrum of AM raw materials and corresponding legal as well as social compliance. Reduced mechanical performance, high cost, lack of modeller availability, etc. are some aspects related to currently available raw materials that restrain full exploitation of AM technology. These issues need serious consideration, especially when fabricating parts for key industries like biomedical, construction, aerospace, automotive, etc. Functionally graded materials and hybrid materials offer some respite to these issues. Still, there is a long way to go before all the AM raw material aspects are fully addressed.

In this chapter, an attempt has been made to cover all the necessary raw material related aspects of AM. The next chapter covers AM design and strategies, i.e. the concept of design for additive manufacturing (DFAM) and its various considerations.

References

1. Rathee, S., Srivastava, M., Maheshwari, S., Kundra, T. K., Siddiquee, A. N., *Friction based additive manufacturing technologies: principles for building in solid state, benefits, limitations, and applications,* 1st ed., 2018, CRC Press, Taylor & Francis Group.
2. Tofail, S. A. M., Koumoulos, E. P., Bandyopadhyay, A., Bose, S., O'Donoghue, L., Charitidis, C., Additive manufacturing: scientific and technological challenges, market uptake and opportunities. *Materials Today,* 2018, **21**(1): pp. 22–37.
3. Bikas, H., Stavropoulos, P., Chryssolouris, G., Additive manufacturing methods and modelling approaches: a critical review. *The International Journal of Advanced Manufacturing Technology,* 2016, **83**(1): pp. 389–405.

4. Mota, R. C. de A. G., da Silva, E. O., de Lima, F., F., de Menezes, L. R., Thiele, A. C. S., 3D printed scaffolds as a new perspective for bone tissue regeneration: literature review. *Materials Sciences and Applications*, 2016, **7**(08): p. 430.
5. Bae, C.-J., Diggs, A. B., Ramachandran, A., Quantification and certification of additive manufacturing materials and processes, in *Additive manufacturing*, J. Zhang and Y.-G. Jung, Editors, 2018, Butterworth-Heinemann, pp. 181–213.
6. Stansbury, J. W., Idacavage, M. J., 3D printing with polymers: challenges among expanding options and opportunities. *Dental Materials*, 2016, **32**(1): pp. 54–64.
7. Wohlers, T., *3D printing and additive manufacturing state of the industry*. Annual Worldwide Progress Report, 2014, Wohlers Associates.
8. Caffrey, T., Wohlers, T., Campbell, I., *Wohlers report 2017: 3D printing and additive manufacturing state of the industry annual worldwide progress report*, 2017, Wohlers Associates.
9. Srivastava M., Rathee, S., Maheshwari, S., Siddiquee, A. N., Kundra, T. K., A review on recent progress in solid state friction based metal additive manufacturing: friction stir additive techniques. *Critical Reviews in Solid State and Materials Sciences*, 2018.
10. DebRoy, T., Wei, H. L., Zuback, J. S., Mukherjee, T., Elmer, J. W., Milewski, J. O., Beese, A. M., Wilson-Heid, A., De, A., Zhang, W., Additive manufacturing of metallic components: process, structure and properties. *Progress in Materials Science*, 2018, **92**(Supplement C): pp. 112–224.
11. Lee, J.-Y., An, J., Chua, C. K., Fundamentals and applications of 3D printing for novel materials. *Applied Materials Today*, 2017, **7**: pp. 120–133.
12. Gosselin, C., Duballet, R., Roux, P., Gaudillière, N., Dirrenberger, J., Morel, P., Large-scale 3D printing of ultra-high performance concrete: a new processing route for architects and builders. *Materials and Design*, 2016, **100**: pp. 102–109.
13. Bourell, D., Kruth, J. P., Leu, M., Levy, G., Rosen, D., Beese, A. M., Clare, A., Materials for additive manufacturing. *CIRP Annals*, 2017, **66**(2): pp. 659–681.
14. Li, R., Liu, J., Shi, Y., Wang, L., Jiang, W., Balling behavior of stainless steel and nickel powder during selective laser melting process. *The International Journal of Advanced Manufacturing Technology*, 2012, **59**(9): p. 1025–1035.
15. King, W. E., Barth, H. D., Castillo, V. M., Gallegos, G. F., Gibbs, J. W., Hahn, D. E., Kamath, C., Rubenchik, A. M., Observation of keyhole-mode laser melting in laser powder-bed fusion additive manufacturing. *Journal of Materials Processing Technology*, 2014, **214**(12): pp. 2915–2925.
16. Li, C., Fu, C. H., Guo, Y. B., Fang, F. Z., Fast Prediction and Validation of Part Distortion in Selective Laser Melting. *Procedia Manufacturing*, 2015, **1**: pp. 355–365.

14

Additive Manufacturing Design and Strategies

14.1 Introduction

There is an ever-increasing demand for tailoring the design process to eliminate the difficulties in manufacturing and minimizing the various associated costs. This has given rise to the ascent of concept of design for manufacturing (DFM). However, DFM capabilities need rethinking owing to the unique nature of AM techniques where there is a demand for customization at different levels. From manufacturing hundreds of person-specific parts as in dentistry, to fabricating highly intricate parts requiring varied design iterations like aircraft components, the demand on AM design strategies is very pressing. AM enables innovative customization abilities, product and process performance enhancement, multifunctional parts, as well as abilities and cost reduction. There are different capabilities of AM which can be categorized as shape, hierarchical, material and functional complexity. This implies that AM techniques can fabricate shapes of any intricacy. This comes from the capacity to fabricate layers of any shape, completely automated process planning, as also the independence of enabling tailored geometries. AM can result in obtaining multi-scaled structures from micro- as well as macrostructural levels resulting in considerable control over microstructural properties by careful control of associated processing parameters. Associated abilities of controlling part nano-, micro-, macro- as well as mesostructures go a long way in giving these technologies unparalleled advantages over conventional design techniques. AM can process materials at different levels and can manufacture completely functional parts and assemblies. Access to inside portions of the parts is the most vital factor responsible for this ability, which in turn provides flexibility to create joints and subparts insertion during the part fabrication process. This in turn reduces much post-processing as well as assembly operations. Material processing can be performed at different points in a variable fashion using AM techniques. Also, composition can be varied abruptly or gradually at any point of the designer's choice. This opens an altogether newer avenue of graded as well as multi-material parts via the AM route in an extremely simple and effective manner. Many interesting applications have also emerged from this particular capability. The immense design freedom offered by AM is its most intriguing characteristic. Applications like significantly reduced assembly requirements and enhanced part consolidation, redesigning, hierarchical structures like cellular material, daily use multi-functional highly customized designs, aircraft components, automotive parts, etc. utilize this particular aspect of AM's unique ability.

This is the second chapter of Section C, which covers design and quality aspects of AM processes. In the previous chapter, all the necessary raw material related aspects of AM have been discussed. This chapter outlines AM design and strategies and covers topics including design for additive manufacturing, AM design tools, design considerations, design strategies and DFAM system details. It concludes the discussion with a summary.

14.2 Design for AM

Design for assembly and manufacturing (DFA/DFM) is the most widely utilized conventional design approach. Though it is seemingly simple to understand, yet its implementation requires a lot of prior knowledge of assembly as well as manufacturing. This in turn requires incorporation of many skills and tools for its successful practice. Industry practices, rules database and consistent research as well as upgrading are the three most important aspects of DFA/DFM. The first aspect includes much professional and smooth communication between design and manufacturing levels for obtaining parts. The second aspect is related to standard design data books for specified principles. The third aspect involves development of specific tools and procedures to evaluate design manufacturability for various specific processes. There have been tremendous efforts in the domain of DFA/DFM because there is a need to understand limitations on various manufacturing processes and subsequent designing with an intention for minimal violation of constraints. There are some ways in which these limitations can be overcome by the path of manufacturing parts using AM. Rules or methods need the flexibility to adapt themselves to the product in order to incorporate the dimension of design freedom offered by the AM route, while simultaneously ensuring non-violation of manufacturing constraints. In DFA, a wide number of repeated iterations is desirable for minimizing the parts/fasteners involved. This in turn tends to increase the cost incurred, so that the increased cost of manufacturing may normally not be compensated by cost saving due to reduced assembly cost.

In design for additive manufacturing (DFAM), it is not necessary to be concerned about the location and orientation of features like overcuts, over-hangs, holes, sprockets, etc. since it has the exemplary uniqueness of fabri-cating a part with all these features without any specific consideration regarding their spatial orientation. For example, if we need to assemble an aircraft component that consists of around 20 subparts, then many opera-tions like stamping, forming, screwing and so on, as well as their detailed understanding, are involved. However, with DFAM, all these are elimi-nated and the part can be obtained as one piece thereby resulting in an inte-grated design that can only be realized using DFAM.

In DFAM, the objective is to maximize product performance, in contrast to the need to reduce operations in DFA/DFM by synthesizing varied shaped/sized parts with hierarchical material compositions. There are several key points that need consideration in DFAM. Intricate geometries can be obtained via the DFAM route without any additional penalty in terms of time or cost. Parts of virtually any complexity can be obtained by the AM route. Customized parts can be obtained directly from three-dimensional CAD data. Realization of a mass customization concept is pos-sible by DFAM as against mass production of conventional techniques. There is a tremendous reduction in assembly requirements in the case of DFAM. Integrated assemblies are possible by this route which offers numerous advantages over conventional approaches. Consolidation of parts with feature integration can be easily obtained in this case. Limita-tions presented by conventional fabrication techniques are also reduced to a great extent.

The issues faced by DFAM are related to the exploration of newer design spheres for exploring advanced and innovative aspects of material and designs. In fact, these can be better understood as opportunities than real challenges; nevertheless they require thorough investigations as well as design creativity to fully explore the avenues created by DFAM.

14.3 AM Design Tools

The 3D CAD file and its corresponding .stl file is the key element of any DFAM problem. However, there are quite a few limitations on the currently available solid modelling software owing to which their complete integra-tion with AM capabilities is not still possible. For example, designing parts with extremely complicated shapes is very difficult and can require a lot of time as well as designer's expertise. Sometimes, there is great difficulty in completely specifying the material composition of parts to be produced. The difficulties in CAD, which is the main AM design tool, can be broadly classed under three heads: (1) difficulty in geometrical representation of parts with numerous features; (2) difficulty in physical representation of meaningful extremely complicated material distribution; and (3) difficulty in property representation of an AM part with continuously varying properties throughout that part.

Another AM design tool is the solid modelling techniques, which include a number of software programs like ProEngineer, SolidWorks, etc. They are quite effective in designing parts with many surfaces and shapes involved. They normally use constructive solid geometry and boundary representation to completely define part topology. However, the memory required with an increase in features is remarkably high since a lot of meaningful data is generated to completely define any given feature. This may lead to lowered processing speeds with an increase in number of surfaces and features. Any solid modelling technique that can be used as a DFAM tool should possess the capability to fully represent features that may range up to thousands for extremely intricate parts. Additionally, it should have the capacity to manage these features of varying dimensions and orders. In general, implicit and multiscale modelling are two CAD categories. This is a serious setback in existing systems since most of the systems have constrained ability to fully define composites, multi-material or graded materials. This also puts a restraint on the design flexibility of the existing systems. To fully utilize varied DFAM capabilities, there are two main domains that need complete attention; these are a mechanism to integrate property–material relationships into geometric modelling data and enabling part synthesis.

There are unique ways in which we can explore the design freedom offered by DFAM principles. Parts can be significantly consolidated, hierarchical structures using meso- and macro-level features can be obtained, newer and innovative products can be obtained via these principles. Consolidating structures and combining them into one lead to numerous advantages over manufacturing parts in single pieces. Difficulties of tooling, assembly, fixtures, fasteners, etc. can be reduced to a great extent. High-end industries including automotive and aerospace greatly benefit from this DFAM capability since there is a considerable engineering saving in terms of money, time, weight and effort. Customized lattices can be imparted to the structures for various functional applications. Intriguing applications of parts made by incorporating DFAM principles are being increasingly applied by the industries and researchers. Multifunctional tailored parts can easily be obtained by utilizing the principles of DFAM.

Figure 14.1 shows a few real-life objects that reflect the immense design creativity imparted by the DFAM approach. Figure 14.2 presents an illustration of two different meso-structures (b and c) obtained by varied spatial orientations of the octet truss cell (a).

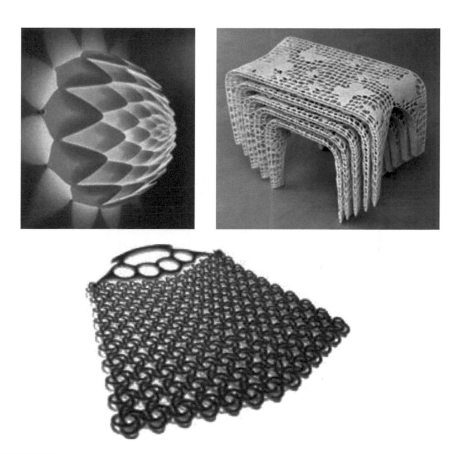

FIGURE 14.1
Design Creativity Freedom in Various Objects Fabricated by AM route. (*Source:* [1]. Reprinted/adapted from Gibson et al., Design for additive manufacturing, in *Additive manufacturing technologies: 3D printing, rapid prototyping, and direct digital manufacturing*, 2015, Springer New York, pp. 399–435, with permission from Springer Nature.

14.4 Design Considerations

There are a wide variety of important design considerations that need to be specifically considered during DFAM. These mainly include the factors affecting the DFAM process, staircase effect in AM parts, resolution obtained, benchmarking considerations, etc. A non-exhaustive list of the common factors that need consideration during DFAM and are applicable to almost all AM processes is presented in Table 14.1, which highlights their geometrical and mechanical effects. If these are judiciously used, then the generic principles of DFAM can be obtained.

TABLE 14.1

Geometrical and Mechanical Effects of Various DFAM Process Parameters

Input parameters	Geometrical effects	Mechanical effects
Feature angle	Larger angle in relation to Z-direction corresponds to lower accuracy and higher roughness on sides	Most processes exhibit anisotropic properties. In addition, mechanical property-angle relationships are not all similar. For example, for powder bed melting AM, fatigue strength and tensile strength may have opposite angular trends
Extrusion/recession features	Recessed features can generally achieve smaller minimum feature sizes. Recessed features may be clogged. Recession and extrusion features are subject to completely different heat transfer conditions during the processes	Recession and extrusion features are subject to completely different heat transfer conditions during the processes
Overhang	Larger overhanging angle or larger overhanging area generally corresponds to more significant loss of accuracy. Down-side facing surfaces of the overhanging features usually have rougher finish	Overhanging features are subject to different heat transfer conditions as the downward heat transfer is restricted
Cross-sectional dimensions	Larger cross-section may indicate larger heat accumulation during process and larger heat sink afterwards, which affects thermal distortion	Larger cross-section may result in more significant in-situ heat treatment effects due to heat accumulation and heat sink effect. For processes that adopt contour-hatch fabrication strategies, different dimensions may result in different combination of deposited properties

Change of cross-sectional area dimension	Similar to cross-sectional effects. Change of dimensions may result in additional stress concentration due to change of heat transfer conditions, which may cause additional thermal distortions	Change of dimensions may result in change of microstructure and properties
Input energy density	Input energy may affect how materials are fabricated. For example, insufficient energy input may cause melting pool instability in powder bed melting AM processes and therefore affect part accuracy and surface finish	There often exists a window of input energy density that results in relatively optimal mechanical properties
Layer thickness	Smaller layer thickness usually corresponds to higher part accuracy at the cost of manufacturing time	For some AM processes, layer thickness indicates thickness of a "laminated composite" structure, therefore is closely associated with the properties. In some processes individual layers serve as boundaries of microstructural grains. Smaller layer thickness may result in better side surface finish which is beneficial to fatigue performance
Material feedstock density	Lower material feedstock could result in more shrinkage during process. For some processes, lower feedstock density also indicates higher process instability, which reduces geometrical accuracy	Higher material feedstock density often corresponds to higher overall mechanical performance due to the reduced amount of internal defects
Preheat	Preheat may cause additional material distortion or loose powder attachment, which reduces accuracy. On the other hand, preheat reduces thermal distortion and therefore could be beneficial to overall part accuracy	Preheat affects thermal gradients in the processes, which in turn affect the microstructure. Preheat could sometimes serve as in-situ heat treatment and alter the microstructure

Source: [2]. Yang et al., Design for additive manufacturing, in *Additive manufacturing of metals: the technology, materials, design and production*, 2017, Springer International, pp. 81–160, reprinted/adapted with permission from Springer Nature.

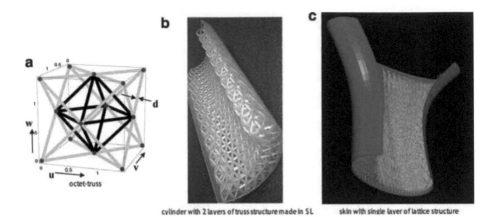

cylinder with 2 layers of truss structure made in SL   skin with single layer of lattice structure

FIGURE 14.2
(a) Octet Truss Unit Cell; (b) and (c) Mesostrucures Obtained from Octet Truss Unit Cell. (*Source:* [1]. Reprinted/adapted from Gibson et al., Design for additive manufacturing, in *Additive manufacturing technologies: 3D printing, rapid prototyping, and direct digital manufacturing*, 2015, Springer New York, pp. 399–435, with permission from Springer Nature.

The staircase effect is an important observation with the AM processes and needs careful consideration during DFAM. This is directly influenced by the thickness of the layer used for fabrication of parts. This effect in various AM parts is shown in Figure 14.3 [2].

Again, resolution of features has huge dependence on the geometrical aspects of the part to be fabricated. Thus, this quality has process dependence.

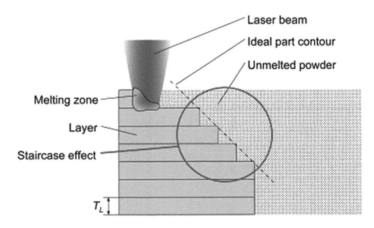

FIGURE 14.3
Illustration of Staircase Effect in Additively Manufactured Parts. (*Source:* [3]. Reprinted from Emmelmann et al., Design for laser additive manufacturing, in *Laser additive manufacturing*, pp. 259–279, Copyright 2017, with permission from Elsevier.)

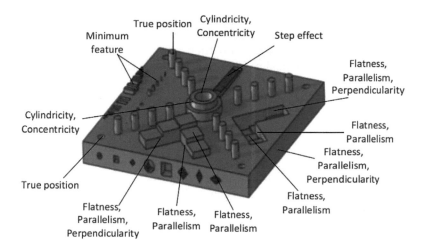

FIGURE 14.4
Standard PBF Benchmark Proposed by NIST. (*Source:* [2]. Reprinted/adapted from Yang et al., Design for additive manufacturing, in *Additive manufacturing of metals: the technology, materials, design and production*, 2017, Springer International, pp. 81–160, with permission from Springer Nature.

There are a few benchmarking techniques to generalize the geometrical features and these can be grouped as mechanical, geometric and process benchmarking. All of these are utilized for evaluating different aspects of design. One standard benchmark proposed by the National Institute of Standards and Technology (NIST) for PBF AM processes is shown in Figure 14.4. The quality judgment is also a key aspect of DFAM and it can be controlled by two different routes, which include in-process and post-process quality control.

14.5 DFAM Design Strategies

Many researchers have covered and presented various important aspects of DFAM [1, 4–6] including but not restricted to design methodology, functionality, cellular materials, compliant systems, production considerations, multi-material and multifunctional parts and so on.

To fully understand the associated DFAM design methodology strategies, there are two main challenges: the creation of a new design space and their subsequent development. One such methodology suggested by Ponche et al. [7, 8] to achieve part functionality, manufacturing and associated physical phenomena consists of three steps which are part orientation, and functional and manufacturing path optimization.

Design for part functionality implies optimizing design for enhancing functionalities, for instance, reducing weights. Here the design targets are given a mathematical definition and then some optimization technique is applied in a predefined design space within constraints. Size (for dimensions), shape (for surfaces) and topological (for material distribution) considerations constitute three main design aspects. Software packages like Abaqus, Optistruct, etc., can be used for optimizing aspects like overall profile, arrangement and connectivity of design elements, etc. pertaining to topological optimization. One example of the topological design aspect is presented in Figure 14.5 where the design domain and corresponding optimized solution are presented for a cargo sling.

Creative designs of intricate 2D and 3D mechanisms can be obtained using DFAM principles. Relative motion of output with respect to input is achieved by designing bending patterns in compliant mechanisms, i.e. bending of structural elements cause relative input–output movement in compliant designs. An example of the compliant mechanism is presented in Figure 14.6. A deployable structure is a variation of compliant structure.

Cellular structures utilize the DFAM intricacy capability and include foams, honeycombs and lattices. They possess appreciably high strength-to-weight ratios, excellent energy absorption, enhanced thermal and acoustic insulating properties, etc. Lattice structures, which are a special class of cellular materials, offer numerous advantages including low weight, high strength and stiffness. Conformal lattices (CLS), where the arrangement of unit cells conform to object surfaces, are one special class of lattice structures. An example of a CLS is presented in Figure 14.7.

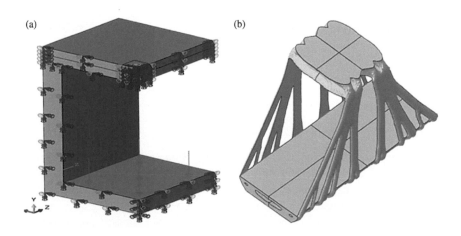

FIGURE 14.5
Topological Design Aspect for Cargo Sling: (a) Design Domain 3'3' 6 m Size with 0.3 m Thick Material; (b) Optimal Solution. (*Source:* [6]. Reproduced with permission from Rosen, Research supporting principles for design for additive manufacturing. *Virtual and Physical Prototyping*, 2014, 9(4): pp. 225–232.)

FIGURE 14.6
Illustration of a Compliant Mechanism. (*Source:* [6]. Reproduced with permission from Rosen, Research supporting principles for design for additive manufacturing. *Virtual and Physical Prototyping*, 2014, 9(4): pp. 225 232.)

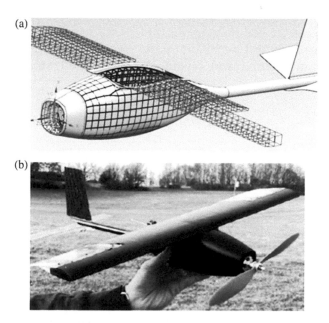

FIGURE 14.7
Hand-Held Unmanned Aerial Vehicle: (a) Design; (b) Object Ready to Fly. (*Source:* [6]. Reproduced with permission from Rosen, Research supporting principles for design for additive manufacturing. *Virtual and Physical Prototyping*, 2014, 9(4): pp. 225–232.)

(a) A biomedical implant   (b) A metal joint part

FIGURE 14.8
Intricate Support Structures Used for Various AM Applications: (a) Biomedical Implant; (b) Metal Joint Part. (*Source:* [2]. Reprinted/adapted from Yang et al., Design for additive manufacturing, in *Additive manufacturing of metals: the technology, materials, design and production*, 2017, Springer International, pp. 81–160, with permission from Springer Nature.

Design of support structures is also an important DFAM aspect. Even when support structures need to be minimized, their utilization is mandatory for various DFAM processes. There are various software packages and algorithms available for software generation. The design and removal require much careful planning; for example, support structures required for biomedical implants and metal joint parts are shown in Figure 14.8. Based upon careful orientation and layout design, support structures can be minimized.

14.6 DFAM System Details

In general, an effective DFAM system should possess the ability to completely answer all the issues related to part design, its material as well manufacturing. This can lead to the requirement of formulating a considerably complex problem that can address the material-process-design-property-manufacturing relationships. One such example is presented for reference in Figure 14.9.

The F42 committee report, associated with ASM International, and the TC 261 committee report, associated with ISO, are mainly related to the standardization of the DFAM approach. These committees have come up with several standard DFAM guidelines. In a general guide issued by them, various key elements that have been suggested include opportunities, issues, trends and limitations in designing for AM systems. Points to be considered during developing products, uses, business, parametric, quality, communication, etc. are covered in the guidebook [6]. This is summarized in Figure 14.10, which shows the related important aspects of DFAM and has been adapted from the initial proposal by the 'European Union project

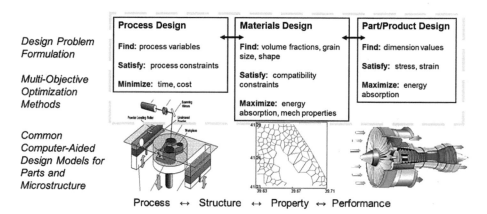

FIGURE 14.9
Material–Process–Design–Property–Manufacturing Relationship for DFAM. (*Source:* [6]. Reproduced with permission from Rosen, Research supporting principles for design for additive manufacturing. *Virtual and Physical Prototyping,* 2014, 9(4): pp. 225–232.)

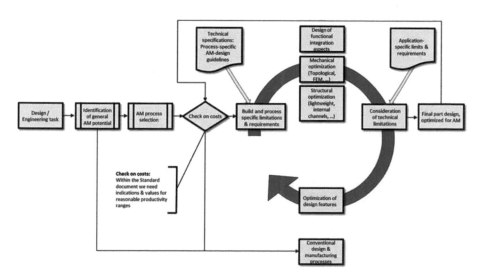

FIGURE 14.10
Example for Design of DFAM. (*Source:* [6]. Reproduced with permission from Rosen, Research supporting principles for design for additive manufacturing. *Virtual and Physical Prototyping,* 2014, 9(4): pp. 225–232.)

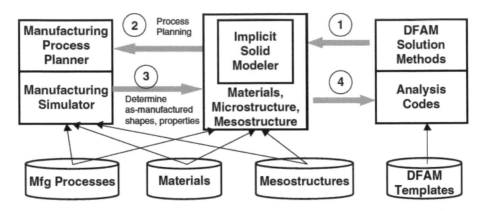

FIGURE 14.11
Basic DFAM Architecture. (*Source:* [1]. Reprinted/adapted from Gibson et al., Design for additive manufacturing, in *Additive manufacturing technologies: 3D printing, rapid prototyping, and direct digital manufacturing,* 2015, Springer New York, pp. 399–435, with permission from Springer Nature.

on standardization of Additive Manufacturing (SASAM)' [6]. Apart from this, many independent researchers have also proposed various set of process specific design rules.

A DFAM system is proposed and its structure is explained through the self-explanatory Figure 14.11. The concepts of implicit, non-manifold as well as parametric modelling is applied to obtain this system.

14.7 Summary

DFAM is a promising and innovative approach. As per Rosen [6], it can be safely concluded that the DFAM objective is: "Maximise product performance through the synthesis of shapes, sizes, hierarchical structures, and material compositions, subject to the capabilities of AM technologies." It holds tremendous opportunities of design and product customization. Other related technologies can be combined with implicit and solid scale modelling for accomplishment of targets in case of DFAM.

Enabling complete exploitation and utilization of various distinctive AM features is feasible only with the aid of fully defined design principles, including but not restricted to geometry, customization, consolidation and final fabrication related aspects. An important observation in this respect is that though design principles corresponding to specific processes have been developed by various individual researchers, a comprehensive overall model for DFAM is still in its conception stage. Creativity, innovation and versatility offered by DFAM need to be fully explored. It can also be concluded that various material related and geometrical design-based aspects of DFAM have been researched in detail but their integration is relatively unexplored by researchers and professionals. This needs to be fully overcome to completely utilize AM.

This is the last chapter of Section C of this book and covers the various AM design and strategies, i.e. concept of DFAM and its various considerations. The next section is Section D, which covers the trends, advancements, applications and conclusion; it has four chapters, 15 to 18.

References

1. Gibson, I., Rosen, D. W., Stucker, B., Design for additive manufacturing, in *Additive manufacturing technologies: 3D printing, rapid prototyping, and direct digital manufacturing*, 2015, Springer, pp. 399–435.
2. Yang, L., Hsu, K., Baughman, B., Godfrey, D., Medina, F., Menon, M., Wiener, S., Design for additive manufacturing, in *Additive manufacturing of metals: the technology, materials, design and production*, 2017, Springer, pp. 81–160.
3. Emmelmann, C., Herzog, D., Kranz, J., Design for laser additive manufacturing, in *Laser additive manufacturing*, M. Brandt, Editor, 2017, Woodhead Publishing, pp. 259–279.
4. Lipson, H., Kurman, M., *Fabricated: the new world of 3D printing*, 2013, John Wiley & Sons.
5. Srivastava, M., *Some studies on layout of generative manufacturing processes for functional components*, 2015, Delhi University, India.
6. Rosen, D. W., Research supporting principles for design for additive manufacturing. *Virtual and Physical Prototyping*, 2014, 9(4): pp. 225–232.
7. Ponche, R., Hascoet, J. Y., Kerbrat, O., Mognol, P., A new global approach to design for additive manufacturing. *Virtual and Physical Prototyping*, 2012, 7(2): pp. 93–105.
8. Ponche, R., Kerbrat, O., Mognol, P., Hascoet, J. Y., A novel methodology of design for additive manufacturing applied to additive laser manufacturing process. *Robotics and Computer-Integrated Manufacturing*, 2014, 30(4): pp. 389–398.

Section D

Trends, Advancements, Applications and Conclusion

15

Hybrid Additive Manufacturing

15.1 Introduction

Additive manufacturing (AM) has evolved over the last 30 years and the research related to AM is continuously ongoing. Since its invention, several AM techniques have evolved, and been patented and commercialized. These techniques can be classified on the basis of different criteria [1]. Several kinds of materials such as polymers, paper laminates, waxes, etc. were initially processed using AM techniques. Subsequently, with the advancements in the AM technology, other materials like metals, composites, ceramics, hybrids, etc. have been introduced. These materials can be broadly classified into metallic and non-metallic types. To date, non-metallic materials are more researched as compared to metallic materials. However, recently, more emphasis has been given to the AM of metallic materials. Several techniques, such as processes based on PBF like EBM, SLS, etc.; processes based on DED like laser consolidation, arc additive manufacturing, etc.; processes based on BJ such as inkjet 3D printing; and processes based on sheet lamination such as LOM, etc. are in common practice for developing metallic components. Most of these techniques, such as EBM, SLS, arc AM, etc., are based on fusion principles. Owing to the molten pool in these techniques the fabricated components are more prone to form discontinuities such as internal porosity, inclusions, and other solidification related defects which lead to poor mechanical properties. In view of these shortcomings in fusion-based MAM, researchers turned towards hybrid MAM and solid state MAM techniques to provide a suitable alternative.

This is the first chapter of Section D, which covers the trends, advancements, applications and conclusion and has four chapters. Section C covered the necessary design and quality aspects of AM processes. This chapter covers solid state hybrid AM techniques, specifically ultrasonic additive manufacturing (UAM) including working principle, benefits and applications, such as embedding of electronic structures into metal matrices, fabrication of reinforced metal matrix composites and metallic laminates; additive manufacturing using cold spraying, including its working principle, advantages, limitations, applications and challenges; friction-based additive manufacturing, including friction deposition-based, friction surfacing-based additive manufacturing, friction stir additive manufacturing, friction assisted seam welding-based additive manufacturing, additive friction stir; conclusion and future scope of hybrid AM techniques. This chapter concludes the discussion with a summary.

15.2 Hybrid AM

Hybrid AM (HAM) processes are basically the manufacturing methods which combine advantages of the cost saving domain of AM with the dimensional accuracy of CNC. The source/means of additive joining varies with the type of hybrid AM process. For example, in UAM, the means of additive joining is ultrasonic vibrations; in welding-based HAM techniques, the means of additive joining is the source of welding (e.g. electric arc, electron beam, etc.), and so on. This means that when a metallic component is manufactured using HAM techniques, initially the component in the form of layers is additively joined via different means as per particular AM technique and then selective machining is performed to provide it with near net shape. The steps of machining are performed either after each additive joining cycle/step/run or after a couple of joining cycles according to suitability of the particular process. The machining step ensures the preparation of the previously built component ready for addition of new layer (to reduce the surface roughness) and also to provide a near net shape. Figure 15.1 presents an overview of machining steps in HAM.

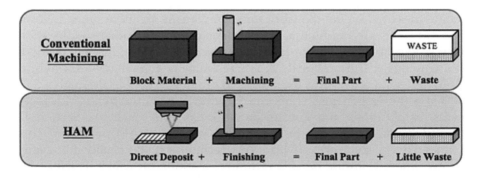

FIGURE 15.1
Basic Schematic of Machining Steps during HAM. (*Source:* [2]. Reprinted from Bandyopadhyay and Heer, Additive manufacturing of multi-material structures. *Materials Science and Engineering: R*, 129: pp. 1–16, Copyright 2018, with permission from Elsevier.)

HAM also works in another way as compared to additive and subtractive combination as discussed in the preceding paragraph. In this method, the machining or subtractive step is performed before the additive step. This use of HAM can be seen in repair type of work. In this way, the worn part/component to be repaired is placed in the HAM system, then first the machining step is performed to restore the original condition and subsequently build formation is done. Such examples can be easily seen in cold spray-based HAM and are discussed later in this chapter.

There are many HAM techniques which are in use. In this chapter, trending solid state HAM techniques are discussed. These techniques mainly include UAM, cold spray-based additive manufacturing (CSAM) and friction-based additive manufacturing (FBAM). These techniques utilize hybrid additive and subtractive steps for fabricating 3D components. As stated above, the additive step in these methods follows particular principles, however, subtractive steps follow selective machining of partly deposited layers using CNC milling machining/operations either integrated with core process set-ups or utilized separately. Being solid state in nature, these methods can produce parts with superior mechanical properties as compared to fusion-based AM techniques where the elevated temperature or melt pool degrades the properties of fabricated components.

15.2.1 Ultrasonic AM

Ultrasonic AM is also called ultrasonic consolidation (UC). This falls in genre of solid state AM techniques which join metallic materials (similar or dissimilar in the form of thin sheets or foils) in layer-by-layer fashion using the principles of ultrasonic welding (USW). It is a hybrid AM technique which combines the additive and subtractive manufacturing principles. In this technique, metallic foils are initially joined using USW and then selective machining is performed using CNC integrated with a UAM system to obtain a near net shape of objects. It was invented and patented by White [3], then commercialized by Solidica Inc. (USA) in 2000, a company founded by Dr. White. This technology is now owned by Fabrisonic Inc. Since its invention, several metals such as aluminium, copper, titanium alloys, etc. have been processed using UAM.

15.2.1.1 Working Principles of UAM

UAM utilizes high frequency ultrasonic vibrations to form a bond between thin sheets/foils to develop 3D solid objects. During its basic working, the metallic sheets (typically 100–300 μm thick) are initially laid out and stacked on a base plate which is generally clamped over a heated anvil. Then, these sheets are compressed under a controlled load or pressure using a rolling tool (known as a sonotrode or horn), as shown in Figure 15.2. Vibration frequency of the rolling sonotrode is around 20 kHz. Vibration direction is transverse to rolling direction of the sonotrode. The amplitude of vibrations generally ranges from 5 to 40 μm. The rolling and vibrating sonotrode grabs the foil owing to its textured surface and both of them vibrate together. Owing to this movement, the interfaces between the foils displace, the asperities of surfaces become flat and the metal-to-metal contact area increases. This leads to softening of the metallic sheets owing to the generation of heat by ultrasonic vibrations. During this process, the plastic flow of materials in the interfacial zone occurs and subsequent recrystallization causes development of newer grains which grow across the interface. As this phenomenon occurs under a controlled load/pressure, atomic bonding between the metallic surfaces starts. The continuous joining of surfaces is provided by the rolling movement of the sonotrode. Figure 15.2 shows the main components of UAM system, while Figure 15.3 shows a working schematic of the UAM process. An illustration of UAM interfacial mechanism is shown in Figure 15.4. Figure 15.5 depicts a UAM machine.

Periodic Machining Operations

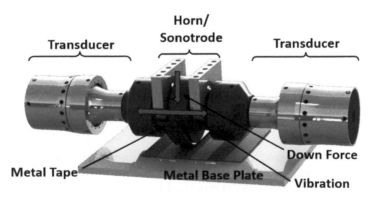

FIGURE 15.2
Main Components of UAM System Showing Additive and Subtractive Steps. (*Source:* Courtesy of Fabrisonic.)

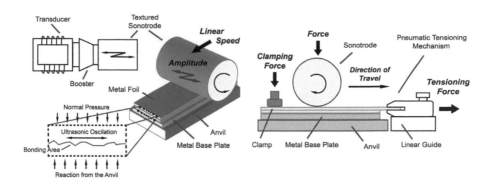

FIGURE 15.3
Schematic Diagram of UAM. (*Source:* [4]. Bournias-Varotsis et al., Ultrasonic additive manufacturing as a form-then-bond process for embedding electronic circuitry into a metal matrix. *Journal of Manufacturing Processes*, 2018, 32: pp. 664–675, under Creative Commons Licence.)

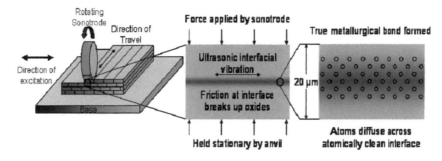

FIGURE 15.4
Illustration of UAM Interface Mechanism. (*Source:* [5]. Johnson, *Interlaminar subgrain refinement in ultrasonic consolidation*, 2008, Loughborough University, UK, under Creative Commons Licence.)

After completion of bonding of a particular layer over the base plate, subsequent layers are added to previously deposited layers/build. Generally, deposition of four foils/sheets is known as one level in UAM. After completion of one level, or as per user's setting, the subtractive step is performed. Generally, a CNC milling tool is utilized to selectively machine the deposited layers to provide a slice contour. The steps of addition of layers/ foils and subtraction/machining are repeated until the desired build dimension and shape is achieved. As the machining step is performed using CNC, the surface finishing and dimensional accuracy is dependent on the CNC except foil thickness or properties which eliminate the stair-stepping effects.

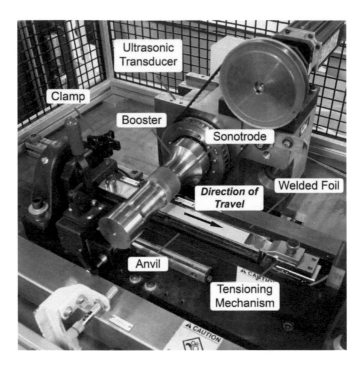

FIGURE 15.5
Alpha 2 UAM Machine. (*Source:* [4]. Bournias-Varotsis et al., Ultrasonic additive manufacturing as a form-then-bond process for embedding electronic circuitry into a metal matrix. *Journal of Manufacturing Processes*, 2018, 32: pp. 664–675, under Creative Commons Licence.)

During the additive step, the joining of foils and their bonding follows the mechanism of USW. USW originated around 60 years ago and is still witnessing consistent growth. It is a solid state joining technique which joins the metallic parts using ultrasonic vibrations under controlled pressure. The basic difference between USW and UAM is that during USW joining of two or more metal plates in different welding positions takes place, while in the latter case multiple sheets/foils are initially joined and then machining operations are performed to shape them into the desired sliced contour.

UAM offers suitability for a wide range of metallic materials for consolidation of similar as well as dissimilar materials. The most common materials used are aluminium alloys. However, components from other materials such as magnesium, copper, titanium, etc. can be suitably fabricated via UAM.

Bonding characteristics/quality are dependent on multiple process parameters and can generally be controlled using three major process parameters:

- Force with which the sonotrode is clamped
- Amplitude with which the sonotrode oscillates at a particular frequency
- Speed of sonotrode.

15.2.1.2 Applications of UAM

Despite several benefits of UAM, the shifting of UAM from laboratories to industrial applications is rare. Current UAM applications are mainly in rapid tooling. Being a low temperature process, UAM has the following current and probable applications:

- Embedding of electronic structures into metal matrices
- Embedding of fibers
- Dissimilar metallic laminates and functionally gradient materials for specific applications
- Fabrication of fiber reinforced metal matrix composites, etc.

Some of these applications are illustrated here via experimental case studies.

15.2.1.2.1 Embedding of Electronic Structures into Metal Matrices

Electronic components and sensors are an essential part of every industry. To enhance the performance of these components, there is a need of encapsulation of these components into a metallic substrate to avoid their degradation owing to corrosion, impact, wear, and so on. UAM offers a more suitable route to develop a protective structure or embedding of electronic components and sensors into metallic substrates as compared to fusion-based AM techniques. This is mainly due to the fact that during UAM no heating is required for bond formation so the electronic components can be easily embedded without any damage, while fusion-based AM techniques generally damage the components due to the involvement of high temperature. Several components such as thermocouples, USB devices, strain sensors, plastic connectors, smart materials, etc. have been successfully embedded into metallic parts using UAM. Figure 15.6 shows some examples of embedded electronic components and sensors into metal parts.

15.2.1.2.2 Fabrication of Reinforced Metal Matrix Composites and Metallic Laminates

UAM allows successful fabrication of MMCs and bonding of metallurgically incompatible materials. This characteristic of UAM makes it suitable for developing functionally graded materials such as metal–metal laminates, fiber reinforced MMCs, etc. Several metal–metal laminates of Al–Al, Al–Cu, Al–Ni, Al–Ti, Ni–Ti and other multi-metallic layer combinations have been developed using UAM. Figure 15.7 shows an Al–Ti laminate

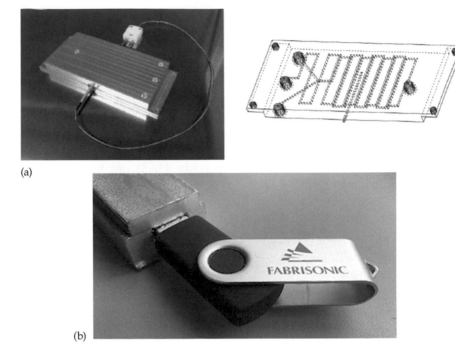

(a)

(b)

FIGURE 15.6
Examples of Embedded Electronic Components and Sensors into Metal Parts: (a) Thermo-couples and Internal Thermal Response from a Heat Plate; (b) USB Connector. (*Source:* Courtesy of Fabrisonic.)

system developed via UAM which is quite difficult to develop using other AM techniques.

An example showing embedding of Ni–Ti shape memory alloy into an aluminium matrix is shown in Figure 15.8. UAM offers a suitable method to embed shape memory alloy at low temperature which enables the shape memory effect to be kept which can be eliminated in fusion-based AM processes.

Another important application of UAM is to produce dissimilar material build/joining without production of harmful intermetallics at the interface. This capability of UAM differentiates it from fusion-based MAM processes. An example is shown in Figure 15.9(a) which is a copper–aluminium lami-nate containing alternate layers of aluminium 6061 alloy and pure copper. Such laminates can be used to optimize the trade-offs among weight and thermal conductivity. Another important example of a build of dissimilar metals fabricated via UAM is shown in Figure 15.9(b) which shows a transi-tion joint of 304 stainless steel and aluminium alloy.

In summary, UAM is a solid state hybrid MAM technique in which ultra-sonic energy is utilized to enable the joining of two metal pieces. UAM is commercialized for tooling application using aluminium alloys.

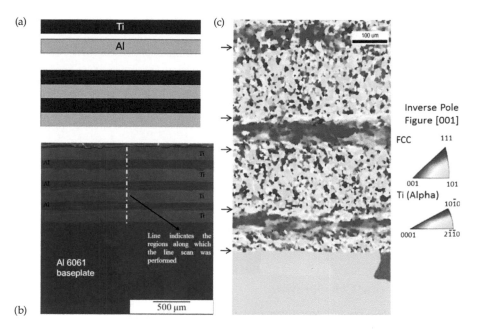

FIGURE 15.7
Showing Al–Ti Laminate Fabricated via UAM: (a) Schematic of Al–Ti Bi-Layer; (b) SEM Image of As-Built Al–Ti Sample; (c) EBSD Image of Ti–Al Bond. Arrows show location of material interfaces. (*Source:* [6]. Reproduced with permission from Wolcott et al., Characterisation of Al–Ti dissimilar material joints fabricated using ultrasonic additive manufacturing. *Science and Technology of Welding and Joining*, 2016, 21(2): pp. 114–123.)

FIGURE 15.8
Cross-Section of Ni–Ti Embedded into Aluminium Using UAM. (*Source:* [7]. Reprinted from Hahnlen and Dapino, NiTi–Al interface strength in ultrasonic additive manufacturing composites. *Composites Part B: Engineering*, 59:, pp. 101–108, Copyright 2014, with permission from Elsevier.)

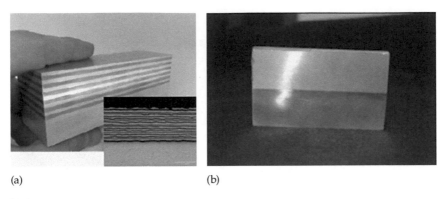

(a)   (b)

FIGURE 15.9
UAM Processed Samples: (a) Laminate of Alternate Layers of Al 6061 and Copper; (b) Transition of Stainless Steel to Al 5052 alloy. (*Source:* Courtesy of Fabrisonic LLC.)

15.2.2 AM Using Cold Spraying

Cold spray-based additive manufacturing (CSAM) is one of the trending solid state hybrid AM techniques. The concept of cold spraying (CS) of metallic materials originated during the 1980s. Papyrin and others [8] initially demonstrated CS applicability at the Russian Academy of Sciences, Novosibirsk. They utilized CS for deposition of a variety of metals, and composites on different types of substrates. This was followed by a grant of one US patent in 1994 and subsequently in 1995 (European patent). Following these, the applicability of CS during last two decades has been continuously increasing. Industrial applications of CS technique include the coating of pure metal and alloys, composites, functional coatings, etc. for better wear resistance, corrosion and high temperature resistance. This is mainly utilized for protective coatings. CS is feasible for all types of pure metals as well as alloys. Exceptionally hard materials are cold sprayed by using a ductile metal binder. Besides coating applications, CS is used as a suitable technique for repair of failed parts. The high velocity and low temperature working conditions makes CS a suitable option for deposits and repair work relative to thermal spraying processes, for example, plasma spraying and high velocity oxy-fuel spray. A comparison of the operating window of CS with other thermal spray techniques is presented in Figure 15.10.

In addition to the coating deposition application, CS has recently been approved as an imperative solid state AM technique in the *Standard Terminology for Additive Manufacturing Technologies* [10]. Also recently the General Electric Company concluded that cold spray-based 3D printing could be utilized to develop 3D components. Although CS is not a new process, however, its application as an additive manufacturing method is quite new.

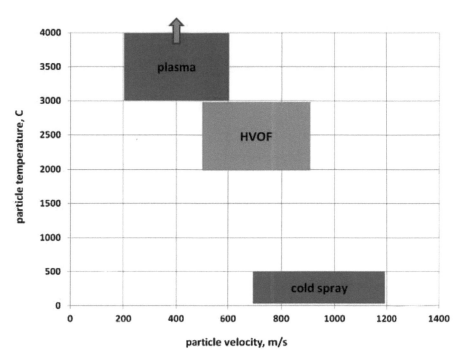

FIGURE 15.10
Operating Window (in Terms of Temperature and Particle Velocity) of Cold Spraying. (*Source:* [9]. Reproduced with permission from Champagne and Helfritch, Critical assessment 11: Structural repairs by cold spray. *Materials Science and Technology*, 2015, 31(6): pp. 627–634.)

15.2.2.1 Working Principles of CSAM

The CS technique finds its basis in the solid state material deposition technique and utilizes a supersonic jet of high temperature compressed gas to accelerate the powder particles for their deposition (using impact) over a substrate. The temperature of the expanded gas stream remains relatively low (100–500°C), and is normally quite less than the powder melting point temperature, which is why it is called cold spray. The material particles are generally in the size range of 10–100 μm in diameter. A CS set-up basically utilizes a compressed gas, a heating system for increasing the temperature of the compressed gas, powder feeding system, supersonic nozzle, robotic arm, as well as some functional arrangement [11]. Figure 15.11 shows a schematic arrangement of a CS system. In its simplest operation, high temperature compressed gas (generally N_2 or He gas stream) expands enroute (converging or diverging) de Laval nozzle to create a supersonic flow. The micron-sized material particles (metals or composites) are exposed to the nozzle inlet and then achieve a high velocity by a supersonic jet; they are then directed towards a substrate or a previously deposited coating

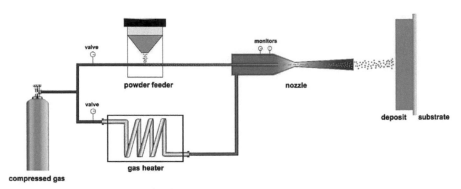

(a) high pressure cold spray system

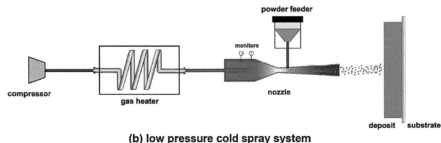

(b) low pressure cold spray system

FIGURE 15.11
Schematic Arrangement of High and Low Pressure CS Systems. (*Source:* [12]. Reprinted from Yin et al., Cold spray additive manufacturing and repair: Fundamentals and applications, *Additive Manufacturing*, 2018, 21: pp. 628–650, with permission from Elsevier.)

forming a bond with the surface owing to impact. The subsequent spray passes increase the thickness of the deposit. For realizing 3D components, machining is needed.

The machining step can be performed in two ways, as proposed by Li et al. [13]. In one method, the machining can be performed during the cold spraying additive process and may be an integral part of the CSAM set-up. In this way, the machining step is performed after deposition of a particular layer of thickness and can be understood as an alternate additive and subtractive process as discussed in Section 15.2. In the other method, the machining step is performed after the completion of deposit thickness to achieve desired dimensions and shape. This can be understood from the illustrative example shown in Figure 15.12, which shows a steel bracket developed via the CS technique over a shaped mandrel substrate at the United Technologies Research Center [14]. After the development of the steel bracket, the substrate was removed and a machining operation was performed over the bracket to achieve the desired finish.

Another example of a CSAM-based component is illustrated in Figure 15.13, which shows an additively manufactured 3D part made of Cu and tool steel with cooling channels from Hermle Maschinenbau Gmbh, Ottobrunn, Germany [15].

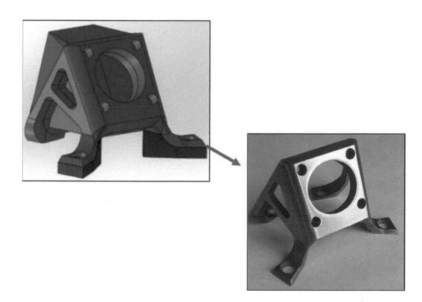

FIGURE 15.12
Steel Bracket Developed over a Substrate via CS; Substrate was Then Removed and Bracket Was Machined to Obtain Desired Shape and Finish. (*Source:* [14]. Reproduced from Nardi, UTRC (United Technologies Research Center) Presentation, 2013.)

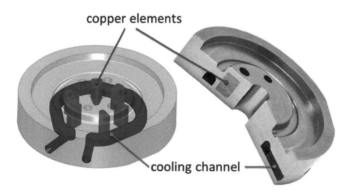

FIGURE 15.13
Additively Manufactured 3D Component Made of Cu and Tool Steel with Cooling Channels via CSAM; Left: Drawing; Right: Fabricated Part. (*Source:* [15]. Reprinted from Assadi et al., Cold spraying: a materials perspective. *Acta Materialia*, 116:, pp. 382–407, Copyright 2016, with permission from Elsevier.)

15.2.2.2 Advantages and Disadvantages of CSAM

In comparison to established fusion-based AM techniques such as EBM, LBM, LMD, etc. CSAM has several unique advantages which mainly include [12, 13]:

- Works in solid state
- Greater flexibility
- Shorter production time
- Suitable for repairing of damaged parts
- Unlimited production volume
- Suitability for production of high reflectivity metals
- Suitable for temperature sensitive substrates
- Suitable for producing composite components.

Thus, CSAM has all the benefits of CS and solid state processing during its layer-by-layer additive manufacturing and repairing/restoration works. As compared to conventional fusion-based AM techniques like SLM, EBM, LMD, CSAM has distinctive benefits which are summarized in Table 15.1.

In addition to the many advantages of CSAM, the parts developed using CSAM suffer from a few disadvantages such as post machining. This is because CSAM deposits have rough surfaces which require post-heat treatment to achieve better mechanical properties.

15.2.2.3 Applications of CSAM

The application of CS can be seen in two ways, i.e. CS in AM and CS in repairing, which are summarized below:

- Repairing of engine blocks of cast iron and nuclear steel containment
- Repairing of worn surfaces
- Repairing of aircraft components
- Fabrication of parts displaying rotational symmetry like cylinder walls, flanges, etc.

A few illustrative examples of repair work using CSAM are shown in Figure 15.14.

An overview of current materials and applications of cold spraying including additive manufacturing for different metals are presented in Table 15.2

TABLE 15.1

Comparison of CSAM with Popular Fusion-Based AM Techniques

Criteria	CSAM	SLM	EBM	LMD
Powder feed mode	Direct deposition	Powder bed	Powder bed	Direct deposition
Feedstock limitations	Difficulty processing high hardness and strength metals	Difficulty processing high reflectivity and poor flowability metals	Unsuitable for non-conductive and low melting-temperature metals	Difficulty with high reflectivity metals
Powder melting	No	Yes	Yes	Yes
Product size	Large	Limited	Limited	Large
Dimensional accuracy	Low	High	High	Medium
Mechanical properties (AF)	Low	High	High	High
Mechanical properties (HT)	High	High	High	High
Production time	Short	Long	Long	Long
Equipment flexibility	High	Low	Low	Low
Suitable for repair	Yes	No	No	Yes

Source: [12]. Yin et al., Cold spray additive manufacturing and repair: fundamentals and applications. *Additive Manufacturing*, 2018, 21: pp. 628–650, with permission.

Note: AF: as-fabricated; HT: heat treatment.

TABLE 15.2

Overview of Current Materials and Applications in Cold Spraying

Material	Application	Sector	Merits
Cu, Ag	Conductive layers High power electronics Sensors Heat transfer layers Corrosion protection	Electronics Automotive Energy Nuclear industries	High chemical homogeneity, no oxidation Similar conductivity as bulk material
Al, Mg alloys	Repair	Aerospace	Low amounts of interface defects Similar strength as bulk material
Al, Ti alloys, Cu, steels	Additive manufacturing Manufacturing of injection/die casting moulds	Aerospace Tooling Polymer industry	Low amounts of interface defects Similar mechanical/fatigue strength as bulk material High deposition rates Complex combination of different materials in one component
Ta, Nb, Cu alloys	Additive manufacturing Sputter targets	Electronics Machinery	High purity, high chemical homogeneity, no oxidation
Ni, Ta, High-Cr steels	Cathodic corrosion protection	Offshore Chemical	Low amounts of interface defects High chemical homogeneity, no oxidation
Zn, Al alloys	Anodic corrosion protection	Offshore Chemical Machinery	High chemical homogeneity, no oxidation Dense coatings
Cu alloys	Tribological performance Cavitation resistance	Maritime technologies Machinery	Low amounts of interface defects Similar mechanical strength as bulk material
Steels ferritic austenitic	Repair Magnetic properties (induction heating)	Energy Consumer goods Machinery	Low amounts of interface defects Similar strength as bulk material Chemical homogeneity, no oxidation
Ni alloys	Hot gas corrosion protection Repair	Energy Aerospace	Low amounts of interface defects Chemical homogeneity, no oxidation
Ti alloys	Biocompatible parts	Medicine	Chemical homogeneity, purity, no oxidation

Source: [15]. Assadi et al., Cold spraying: a materials perspective. *Acta Materialia*, 2016, 116: pp. 382–407.

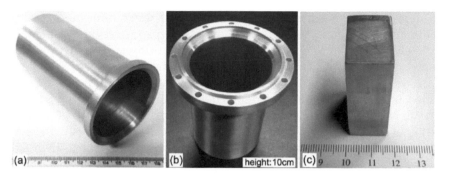

FIGURE 15.14
Components Developed via CSAM Using CS as Additive Process and Different Machining Operations as Subtractive Operations: (a) Al Alloy Tube; (b) Al Alloy Flange; (c) Copper Cuboid. (*Source:* [12]. Reproduced with permission from Yin et al., Cold spray additive manufacturing and repair: fundamentals and applications. *Additive Manufacturing*, 2018, 21: pp. 628–650.)

15.2.2.4 Challenges of CSAM

CSAM has great potential in fabricating newer products as well as in repairing damaged parts. Numerous parts are developed and/or repaired using CSAM technology. Despite several CSAM benefits and its increasing trend in AM and repairing, it is still facing some prominent challenges:

- *Post machining:* CSAM products need a post machining operation which should be reduced/eliminated. This issue should be explored in future research on CSAM.

- *Low ductility:* CSAM products exhibit low ductility. Further, strengthening mechanisms like heat treatments need extensive research.

- *Processing parameters:* CSAM processing parameters such as spraying or powder related parameters are crucial for formation of sound CSAM products. Thus, a thorough understanding of these parameters needs further research.

- *Materials:* Currently CSAM is mainly used for a few metals such as aluminium, copper and so on. Further investigation of other materials like titanium alloys, stainless steel, etc. is needed, which will open newer avenues of CSAM applications in aerospace and other engineering sectors.

15.2.3 Friction-Based AM

Friction-based additive manufacturing (FBAMs) processes are gaining popularity in the AM of metallic components owing to their solid state nature and grain refinement mechanisms [1, 16]. These techniques basically utilize the layer-by-layer additive principle of AM and the joining mechanism of frictional heating and subsequent joining. These processes have capabilities to develop 3D components from a wide range of metallic materials, composites, functionally graded materials, etc. Currently, seven FBAM processes are reported [1]: AM based on rotary friction welding (RFW), linear friction welding (LFW), friction deposition (FD), friction surfacing (FS), friction stir additive manufacturing (FSAM), additive friction stir (AFS) and friction assisted seam welding (FASW). All of these techniques work in the solid state to additively manufacture 3D components. Owing to their solid state nature, these techniques have numerous advantages over fusion-based AM techniques; they are summarized below [1]:

- Environmentally friendly fabrication techniques
- Utilize comparatively lesser energy
- Suitable for fabrication of dissimilar materials build
- Capability to fabricate fine-grained microstructures
- Structural efficacy
- Less distortion
- High reproducibility rate
- Low porosity and solidification related defects

and so on.

Despite its many advantages, FBAM technology is in its infancy and countable milestones need to be achieved in this direction. The major reason behind this lack of growth is the involvement of two different domains, namely the additive manufacturing and friction-based processes. Less available literature and patent technology are also other major reasons for their restricted growth. A timeline of the major milestones achieved with regard to FBAM is presented in Table 15.3.

Brief details of various FBAM processes on the basis of their working principles, strengths and challenges are summarized in Table 15.4. For more details, readers are encouraged to refer the book by Rathee et al. [1].

Among these seven processes, two prominent FBAMs (friction stir additive manufacturing and additive friction stir) are discussed in the next two subsections.

15.2.3.1 Friction Stir AM

Friction stir additive manufacturing (FSAM) is based upon the friction stir welding/processing principle [1, 16]. In this technique, 3D component fabrication is achieved by initial joining of sheets/layers and then performing a selective machining operation. The additive joining mechanism follows the principle of friction stir welding/processing (FSW/P), while the selective

TABLE 15.3

Timeline of FBAMs

S. No.	Year	Remarkable Progress
1	2002	White et al. [17] patented friction joining for AM
2	2005	Thomas et al. [18] stated that additive FSW method (i.e. near net shape prototyping method) is under development at The Welding Institute
3	2006	Airbus [19] presented maiden report stating that FSW/P can be used for AM of components
4	2007	Threadgill and Russell [20] demonstrated rotary and linear friction welding as suitable option for MAM
5	2011	Dilip et al. [21] introduced a novel friction-based AM process and termed it as friction deposition
6	2012	Boeing [22] showed that FSAM can achieve component fabrication via low production and less waste
7	2012	Dilip et al. [23] showed AM as probable route for friction welding and friction deposition processes
8	2013	Dilip et al. [24] demonstrated probable way of friction surfacing for AM
9	2013	Kandasamy et al. [25, 26] proposed additive friction stir process as a suitable route to fabricate defect free aluminium and magnesium components
10	2015	Palanivel et al. [27] used FSAM for microstructural control of magnesium-based alloys
11	2015	Withers [28] suggested FSAM as a potential route for high performing structural applications
12	2015	Palanivel et al. [29] demonstrated FSAM as a route to fabricate high performance Al and magnesium-based alloys
13	2014–2016	NASA project was completed on additive friction stir (AFS) technology which projected multiple advantages for aircraft, space and commercial industries specially in terms of customization, achievement of wrought microstructures, low costs, etc.
14	2015	Calvert presented thesis which intended to study microstructures and mechanical properties of components fabricated via AFS [30]
15	2016	Yuqing et al. [31] presented microstructure, formation characteristics and mechanical performance of Al-based FSAM process
16	2016	Kumar Kandasamy [32] granted patent on AFS
17	2017	Palanivel et al. [33] published a review on friction-based processes including FSAM by projecting it as a technique which builds without melting
18	2017	Rivera et al. [34] reported fabrication of Inconel 625 using AFS
19	2018	Rathee et al. [1] published book on FBAMs
20	2018	Srivastava et al. [16] published review article on friction-based AM techniques

Source: [1, 16]. Rathee et al., *Friction based additive manufacturing technologies: principles for building in solid state, benefits, limitations, and applications,* 1st ed., 2018, CRC Press, Taylor & Francis Group; and Srivastava et al., A review on recent progress in solid state friction based metal additive manufacturing: friction stir additive techniques. *Critical Reviews in Solid State and Materials Sciences,* 2018: pp. 1–33.

TABLE 15.4

Details of Different FBAM Processes

S. No.	FBAM	Working principle	Strength	Challenges	Factors affecting microstructures
1	RFW	One part is kept at rest and other rotates at given speed. RFW has two variants, namely direct drive RF and inertia based RF	(a) Large dimension structures can be built (b) Deposition time independent of individual part dimension (c) Suitable specially for assembling small parts into bigger ones (d) Fine granular interfacial layers (e) Properties equivalent to parent material	(a) Can only be used for round shaped symmetrical parts like disk, tube, cylinder in its present form (b) There are four zones in each layer, i.e, thermo-mechanical, base metal, HAZ and weld interface, which amounts to in-homogeneity (c) Suitable only for layer thickness less than 10mm.	(a) Speed of rotation (b) Upsetting force (c) Joining times (d) Frictional forces (e) Burn-off lengths
2	FD	A stationary part is brought into contact with a rod that is consumed and attached to a spindle undergoing rotational movement. This results in plasticization and subsequent flowing of rod material to form layer on stationary part. It is a modification of RFW	(a) Build structure has no discrete zones which implies perfect weld structure (b) Components have homogeneous microstructures (c) Produces better microstructures than fusion AM processes	(a) All layers are sensitized except top ones (b) Only those parts which can sustain buckling and deformative forces can be fabricated (c) Inadequacy in bonding at edges	(a) Speed of rotation (b) Frictional forces (c) Depositional rates (d) Burn-off lengths
3	FS	A rotating mechatrode presses against substrate due to axial force. Material is then bonded to base owing to diffusion at the onset of visco-plasticity behavior. This further amounts to material consolidation and layer thickening with dwell. This is followed by forward traversing of mechatrode to deposit entire layer	(a) Can be utilized for multiple shapes (b) Tracks are completely consolidated to layered forms (c) Achievable mechanical properties are at par with wrought phase characteristics (d) High deposition rates	(a) Tracks reflect incomplete mutual consolidation (b) Positioning of mechatrode is critical (c) Oxide layers formation on surfaces (d) Unevenly deposited edges	(a) Mechatrode positioning (b) Base part thickness (c) Rod geometry (d) Speed of rotation (e) Speed of traverse

4	LFW	Friction between two surfaces is generated by rubbing surfaces during loading. Here one part is made to undergo oscillations while keeping second part fixed. Material consolidates in four stages, i.e. contact, plasticization, softening and extrusion and forging. Process is repeated to obtain multiple layers	(a) Specially popular for blisk manufacturing in aeroplanes (b) Oscillations can be customized to control buckling and axial loading characteristics (c) Can be utilized for manufacturing of complex geometries	(a) Four zones present in final weld similar to RFW though HAZ lesser thick in LFW (b) Process capabilities limited by oscillatory method applied (c) Huge initial investments (d) Process extremely noisy	(a) Oscillatory frequency (b) Oscillatory amplitudes (c) Axial loading magnitudes (d) Frictional pressure (e) Forge process characteristics
5	FSAM	Explained in Section 15.2.3.1	(a) Energy efficiency structures fabricated (b) Higher production rates (c) Reduced material wastage (d) Higher performance structures (e) High microstructural control (f) Greater tensile strengths and ductility (g) Properties better than wrought structures	(a) Material clamping is required (b) Tool wear rates appreciably higher (c) Microstructure in-homogeneity (d) Machine dependence	(a) Tool parameters (b) Backing-up plate (c) Fixtures to cool (d) Speeds of rotation (e) Forging forces (f) Traverse velocities (g) Plate dimensions (h) Thermal cycles and straining rates
6	AFS	Explained in Section 15.2.3.2	(a) Starting material can be in powdered form (b) Absence of solidification defects (c) Feed material consolidation occurs within tool (e) Almost isotropic properties obtained	(a) Special purpose equipment required (b) Process slightly complex (c) Machine parameter dependence	(a) Geometrical features of tool (b) Spindle speeds (c) Traverse speeds (d) Feeding rates for fillers (e) Interfering layers

Source: [1, 16]. Rathee et al., *Friction based additive manufacturing technologies: principles for building in solid state, benefits, limitations, and applications,* 1st ed., 2018, CRC Press, Taylor & Francis Group; and Srivastava et al., A review on recent progress in solid state friction based metal additive manufacturing: friction stir additive techniques. *Critical Reviews in Solid State and Materials Sciences,* 2018: pp. 1–33.

machining step follows the mechanism of machining (such as CNC machining) [33, 35, 36]. A schematic arrangement of FSAM is shown in Figure 15.15.
FSAM follows simple steps which are summarized below:

- First, two layers/sheets (placed one over other) are clamped on a work fixture, then friction stir lap welding (FSLW) is performed.
- A machining operation is performed to remove the marks of FSLW and to prepare the upper surface of additively joined sheets and also to selectively provide the flatness.
- The next layer is placed over the finished top surface of previous layers and the same steps are repeated till the final build is obtained.

These systematic steps are illustrated in Figure 15.16.

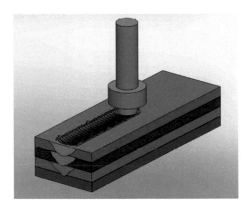

FIGURE 15.15
Schematic Arrangement of FSAM. (*Source:* [1]. Reproduced with permission from Rathee et al., *Friction based additive manufacturing technologies: principles for building in solid state, benefits, limitations, and applications*, 1st ed., 2018, CRC Press, Taylor & Francis Group.)

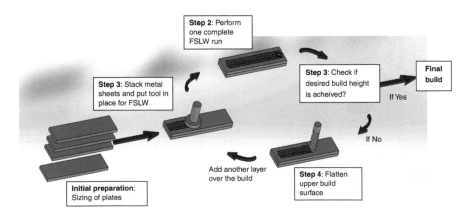

FIGURE 15.16
Steps Utilized in FSAM. (*Source:* [1]. Reproduced with permission from Rathee et al., *Friction based additive manufacturing technologies: principles for building in solid state, benefits, limitations, and applications*, 1st ed., 2018, CRC Press, Taylor & Francis Group.)

15.2.3.2 Additive Friction Stir

Additive friction stir (AFS), also known as MELD, is a kind of solid state AM technique [37]. It is purely additive in nature as compared to other friction-based AM techniques. During AFS, material is added in the form of layers deposition in solid state using principles of FSW/P. Material is fed in powdered form as well as solid rod via a rotating hollow tool/shoulder. The rotational movement of the hollow tool generates heat owing to the friction between the contact interfaces of tool and substrate. Owing to this heat and subsequent plastic deformation at the interfaces, a bond between softened plasticized feedstock material and substrate occurs. When the substrate is allowed to traverse in such conditions, deposition of a single layer takes place over the substrate. The schematic arrangement of AFS is shown in Figure 15.17. To develop 3D objects (layer-by-layer deposition) via AFS, tool height is adjusted to deposit the subsequent layer over previously deposited layers and the same steps are repeated until the desired height of build is achieved.

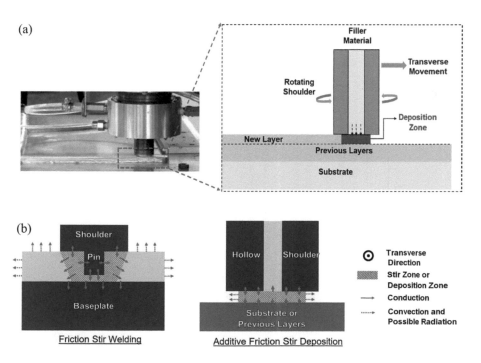

FIGURE 15.17
Schematic Arrangement of AFS/MELD: (a) Rotating Tool (Left, Real Time Image), a Schematic Showing AFS Process Mechanism (Right); (b) Heat Flow in FSW and AFS Deposited Material. (*Source:* [37]. Reprinted from Yu et al., Non-beam-based metal additive manufacturing enabled by additive friction stir deposition. *Scripta Materialia*, 153: pp. 122–130, Copyright 2018, with permission from Elsevier.)

TABLE 15.5

Current and Potential Applications of FBAM Processes

Process	Present applications	Probable applications
Rotary friction welding [23]	Two shafts (similar and dissimilar cross-sections), shaft-flange, two tubes, tube-flanges, etc. in tools, valves, automotive components	Multi-material components like metal forming roll, components with fully enclosed cavity, honeycomb structures, etc.
Linear friction welding [38]	Manufacturing and repair of aircraft blisks, automobile braking blocks, large sized pipes and other components, etc.	Gradient structures fabrication at large scale, airframe brackets, etc.
Friction deposition [39]	Development of ferrous and non-ferrous metal deposits, metal–metal composites, etc.	Injection molding dies and tooling, parts with fully enclosed cavity, circular parts with multi-material composition
Friction surfacing [24]	Coatings for high wear and corrosion applications, repair applications	3D multi-material, in aircraft frames such as rib-on-plate structures, repairing of dies, etc.
Friction stir additive manufacturing [27]	Producing structural components from Al, Mg alloys, preforms fabrication	Functionally gradient materials, stringer in aerospace fuselage, etc.
Friction assisted (lap) seam welding [40]	Variety of industrial welding applications in automotive, aircraft, marine, nuclear and processing industries, etc.	Cladding and AM of similar as well as dissimilar materials, producing high-strength seam welds, AM of wide range of materials, etc.
Additive friction stir [30, 34]	Fabrication of high-strength ultrafine grained magnesium alloys	Coating of shaft journals, airframe structures like bulkheads and stiffeners, etc.

Source: [1, 16]. Rathee et al, *Friction based additive manufacturing technologies: principles for building in solid state, benefits, limitations, and applications,* 1st ed., 2018, CRC Press, Taylor & Francis Group; and Srivastava et al., A review on recent progress in solid state friction based metal additive manufacturing: friction stir additive techniques. *Critical Reviews in Solid State and Materials Sciences,* 2018: pp. 1–33.

AFS produces good metallurgical bonding between deposited coating/ material and substrate and is free from the defects generally occurring in fusion-based AM techniques such as porosity, hot cracking, etc. Parts produced via AFS exhibit fine grained wrought microstructures with superior mechanical properties.

15.2.3.3 Applications of FBAM

Owing to the solid state nature and capability to develop wrought microstructures, F BAM processes have their applications varying from development of new 3D metallic components (simple as well complex shape) to repair of worn-out parts from a wide range of metallic materials. The main applications (which are in current use) of FBAM techniques and their probable applications are presented in Table 15.5.

15.2.4 Comparison between UAM, FSAM and AFS

A comparison of three hybrid solid state AM techniques, i.e. UAM, FSAM and AFS, is presented in Table 15.6.

15.3 Summary

HAM allows the benefits of cost saving and freedom of design principles of AM and dimensional accuracy of CNC machining to be taken. Despite several benefits, HAM possesses a few limitations. The most common limitations from design to final object include the time requirement for changing of tools, in-process and post process machining, pre-design considerations, limited

TABLE 15.6

Comparison of UAM, FSAM and AFS

Process	Ultrasonic additive manufacturing	Friction stir additive manufacturing	Additive friction stir deposition
ASTM classification	Sheet lamination	Sheet lamination	N/A
Hybrid process	Yes	Yes	No
Resolution limiting factor	Subtractive process	Subtractive process	Tool geometry
Temperature	Relatively low	Relatively high	Relatively high
Microstructure	Similar to pre-processed	Refined, equiaxed in the stir zone only	Refined, equiaxed

Source: [37]. Yu et al., Non-beam-based metal additive manufacturing enabled by additive friction stir deposition. *Scripta Materialia*, 2018, 153: pp. 122–130.

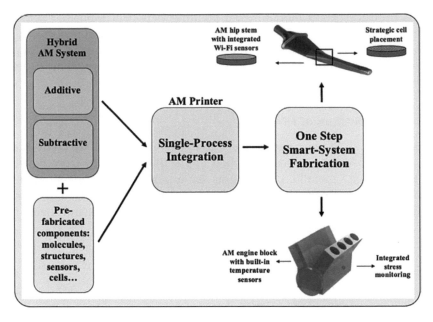

FIGURE 15.18
Futuristic Approach of HAM Systems: Where Pre-Fabricated Parts are Integrated with HAM System to Develop Smart Systems in One Manufacturing Step. (*Source:* [2]. Reprinted from Bandyopadhyay and Heer, Additive manufacturing of multi-material structures. *Materials Science and Engineering: R*, 129: pp. 1–16, Copyright 2018, with permission from Elsevier.)

feasibility for overhanging structures, etc. Once the limitations of HAM are addressed, industries such as automotive, aerospace, defense, medical, and so on will benefit greatly from the commercialization of these techniques. The intriguing application of HAM in the area of repair work is appealing to these industries in terms of cost savings by eliminating replacement costs. The future scope and aspects of HAM system is beautifully described by Figure 15.18.

In this chapter, readers have been introduced to all the latest trends and variants of AM. A detailed discussion of the various AM applications in different sectors is provided in the next chapter of Section D.

References

1. Rathee, S., Srivastava, M., Maheshwari, S., Kundra, T. K., Siddiquee, A. N., *Friction based additive manufacturing technologies: principles for building in solid state, benefits, limitations, and applications*, 1st ed., 2018, CRC Press, Taylor & Francis Group.

2. Bandyopadhyay, A., Heer, B., Additive manufacturing of multi-material structures. *Materials Science and Engineering: R: Reports*, 2018, **129**: pp. 1–16.

3. White, D., *Ultrasonic object consolidation*, U. Grant, Editor, 2003, Solidica.

4. Bournias-Varotsis, A. B. V., Friel, J. R., Harris, R. A., Engstrøm, D. S., Ultrasonic additive manufacturing as a form-then-bond process for embedding electronic circuitry into a metal matrix. *Journal of Manufacturing Processes*, 2018, **32**: pp. 664–675.

5. Johnson, K. E., *Interlaminar subgrain refinement in ultrasonic consolidation*, 2008, Loughborough University, UK.

6. Wolcott, P. J., Sridharan, N., Babu, S. S., Miriyev, A., Frage, N., Dapino, M. J., Characterisation of Al–Ti dissimilar material joints fabricated using ultrasonic additive manufacturing. *Science and Technology of Welding and Joining*, 2016, **21**(2): pp. 114–123.

7. Hahnlen, R., Dapino, M. J., NiTi–Al interface strength in ultrasonic additive manufacturing composites. *Composites Part B: Engineering*, 2014, **59**: pp. 101–108.

8. Papyrin, A., Cold spray technology. *Advanced materials and processes*, 2001, **159**(9): pp. 49–51.

9. Champagne, V., Helfritch, D., Critical assessment 11: Structural repairs by cold spray. *Materials Science and Technology*, 2015, **31**(6): pp. 627–634.

10. ASTM F2792–12a, *Standard terminology for additive manufacturing technologies*, 2012.

11. Li, W., Assadi, H., Gaertner, F., Yin, S., A Review of Advanced Composite and Nanostructured Coatings by Solid-State Cold Spraying Process. *Critical Reviews in Solid State and Materials Sciences*, 2018: pp. 1–48.

12. Yin, S., Cavaliere, P., Aldwell, B., Jenkins, R., Liao, H., Li, W., Lupoi, R., Cold spray additive manufacturing and repair: fundamentals and applications. *Additive Manufacturing*, 2018, **21**: pp. 628–650.

13. Li, W., Yang, K., Yin, S., Yang, X., Xu, Y., Lupoi, R., Solid-state additive manufacturing and repairing by cold spraying: A review. *Journal of Materials Science and Technology*, 2018, **34**(3): pp. 440–457.

14. Nardi, A., UTRC (United Technologies Research Center) Presentation, 2013.

15. Assadi, H., Kreye, H., Gärtner, F., Klassen, T., Cold spraying: a materials perspective. *Acta Materialia*, 2016, **116**: pp. 382–407.

16. Srivastava, M., Rathee, S., Maheshwari, S., Siddiquee, A. N., Kundra, T. K., A review on recent progress in solid state friction based metal additive manufacturing: friction stir additive techniques. *Critical Reviews in Solid State and Materials Sciences*, 2018: pp. 1–33.

17. White, D., *Object consolidation employing friction joining*, U.S. Patent 6457629B1, October 1, 2002.

18. Thomas, W. M., Norris, I. M., Staines, D. G., Watts, E. R., *Friction stir welding: process developments and variant techniques*, SME Summit, 2005, Oconomowoc, Milwaukee, USA.

19. Lequeu, P. H., Muzzolini, R., Ehrstrom, J. C., Bron, F., Maziarz, R, *High performance friction stir welded structures using advanced alloys*, Aeromat Conference, 2006, Seattle, WA.

20. Threadgill, P. L., Russell, M. J., *Friction welding of near net shape preforms in Ti-6Al-4V*, 11th World Conference on Titanium (JIMIC-5), 2007, Kyoto, Japan.

21. Dilip, J. J. S., Kalid, R. H., Janaki Ram, G. D., A new additive manufacturing process based on friction deposition. *Transactions of the Indian Institute of Metals*, 2011, **64**(1): p. 27.

22. Baumann, J. A., *Technical report on production of energy efficient preform structures*, 2012, The Boeing Company, Huntington Beach, CA.

23. Dilip, J. J. S., Janaki Ram, G. D., Stucker, B. E., Additive manufacturing with friction welding and friction deposition processes. *International Journal of Rapid Manufacturing*, 2012, **3**(1): pp. 56–69.

24. Dilip, J. J. S., Babu, S., Varadha Rajan, S., Rafi, K. H., Janaki Ram, G. D., Stucker, B. E., Use of friction surfacing for additive manufacturing. *Materials and Manufacturing Processes*, 2013, **28**(2): pp. 189–194.

25. Kandasamy, K., Renaghan, L. E., Calvert, J. R., Schultz, J. P. Additive friction stir deposition of WE43 and AZ91 magnesium alloys: microstructural and mechanical characterization, in *International conference, Powder metallurgy and particulate materials*, 2013. Chicago, IL.

26. Kandasamy, K., Renaghan, L., Calvert, J., Creehan, K., Schultz, J., Solid-state additive manufacturing of aluminum and magnesium alloys, in *Materials science and technology conference and exhibition 2013: (MS&T'13)*, 2013, Montreal, Quebec, Canada.

27. Palanivel, S., Nelaturu, P., Glass, B., Mishra, R. S., Friction stir additive manufacturing for high structural performance through microstructural control in an Mg based WE43 alloy. *Materials and Design (1980–2015)*, 2015, **65**: pp. 934–952.

28. James Withers, R. S. M., *Friction stir additive manufacturing as a potential route to achieve high performing structures*, 2015, US DOE Workshop on Advanced Methods for Manufacturing (AMM), University of North Texas.

29. Palanivel, S., Sidhar, H., Mishra, R. S., Friction stir additive manufacturing: route to high structural performance. *Journal of the Minerals, Metals and Materials Society*, 2015, **67**(3): pp. 616–621.

30. Calvert, J. R., Microstructure and mechanical properties of WE43 alloy produced via additive friction stir technology, in *Materials Science and Engineering*, 2015, Virginia Polytechnic Institute and State University.

31. Yuqing, M. L., Ke C., Huang F., Liu Q. L., Formation characteristic, microstructure, and mechanical performances of aluminum-based components by friction stir additive manufacturing. *The International Journal of Advanced Manufacturing Technology*, 2016, **83**(9): pp. 1637–1647.

32. Kandasamy, K., *Solid state joining using additive friction stir processing, U.S. Patent US9511445B2*, December 6, 2016.

33. Palanivel, S., Mishra, R. S., Building without melting: a short review of friction-based additive manufacturing techniques. *International Journal of Additive and Subtractive Materials Manufacturing*, 2017, **1**(1): pp. 82–103.

34. Rivera, O. G., Allison, P. G., Jordon, J. B., Rodriguez, O. L., Brewer, L. N., McClelland, Z., Whittington, W. R., Francis, D., Su, J., Martens, R. L., Hardwick, N., Microstructures and mechanical behavior of Inconel 625 fabricated by solid-state additive manufacturing. *Materials Science and Engineering: A*, 2017, **694**(Supplement C): pp. 1–9.

35. Rathee, S., Maheshwari, S., Noor Siddiquee, A., Srivastava, M., A review of recent progress in solid state fabrication of composites and functionally graded systems via friction stir processing. *Critical Reviews in Solid State and Materials Sciences*, 2018, **43**(4): pp. 334–366.

36. Rathee, S., Maheshwari, S., Noor Siddiquee, A., Issues and strategies in composite fabrication via friction stir processing: a review. *Materials and Manufacturing Processes*, 2018, **33**(3): pp. 239–261.
37. Yu, H. Z., Jones, M. E., Brady, G. W., Griffiths, R. J., Garcia, D., Rauch, H. A., Cox, C. D., Hardwick, N., Non-beam-based metal additive manufacturing enabled by additive friction stir deposition. *Scripta Materialia*, 2018, **153**: pp. 122–130.
38. Slattery, K. T., Young, K. A., *Structural assemblies and preforms therefor formed by friction welding*, U.S. Patent *US7854363B2*, November 13, 2008.
39. Dilip, J. J. S., Janaki Ram, G. D., Microstructures and properties of friction freeform fabricated borated stainless steel. *Journal of Materials Engineering and Performance*, 2013, **22**(10): p. 3034–3042.
40. Kalvala, P. R., Akram, J., Misra, M., Friction assisted solid state lap seam welding and additive manufacturing method. *Defence Technology*, 2016, **12**(1): pp. 16–24.

16

Additive Manufacturing Applications

16.1 Introduction

Additive manufacturing (AM) processes constitute an important class of manufacturing technology advancement. There is a huge spectrum of applications of additively manufactured parts which can chiefly be attributed to the development and improvement of AM processes. AM is particularly suitable for applications in producing complex shaped geometrical components as compared to conventional manufacturing techniques. From their initial applications as visualization tools, the innovation of AM parts has witnessed tremendous growth and diversity. The applications of AM processes have almost reached each industrial sector, which mainly includes aerospace, defense, marine, automotive, medical, retail sector, etc.

This is the second chapter of Section D, which covers the trends, advancements, applications and conclusion. The latest AM trends, advancements and variants have been discussed in the previous chapter. This chapter outlines AM applications, especially with respect to their as visualization tools, in the aerospace, automotive, medical, construction industries, and retail applications; it concludes the discussion with a summary.

16.2 Application of AM Parts as Visualization Tools

It is a matter of common knowledge that models are a far better tool to ease new product development. Making assembly drawings for manufacturing parts is a cumbersome and time consuming but economical process. However, model creation for design validation is almost always mandatory. This was also the initial application of AM. In general, models are necessary as a quick source for form, fit and function (3F) information for performance enhancement of the AM part. Form implies inspecting the shape and general need of given design. Fit implies closely agreeing with predefined dimensional tolerances and can mainly be attributed to improvement in process efficiency. Function is related to the performance evaluation of AM parts and is associated with the enhanced material properties.

16.3 AM Applications in Aerospace

One of the highly promising areas of AM is the aerospace industry. The ability of AM to manufacture lightweight parts is the most appealing advantage of AM for the aerospace industry. The components/parts utilized in the aerospace industry manufactured via various techniques should match the absolute target properties to ensure safety. Aerospace components are generally complex in geometrical shape and difficult to manufacture, as well as sometimes costly, via conventional manufacturing techniques. Titanium alloys, aluminium alloys, nickel super alloys, special grades of steel, etc. are commonly utilized materials for manufacturing components for aerospace applications. Manufacturing of these parts fabricated from advanced materials needs care especially, with respect to complex geometrical shapes [1]. In addition, the quantity of parts/components to be produced for aerospace application is small (low volume), there can be a maximum production run of 1000 parts. Owing to the primary aspect of safety, cost efficiency, etc. AM technology is a good choice for manufacturing aerospace components. The following characteristics of components used in aerospace make them suitable candidates to be processed via AM [2]:

- *Hard to manufacture/machine as also excessive buy-to-fly ratio:* As described above, aerospace components are generally made of advanced materials. These are difficult to machine and manufacture, time consuming and costly also. Apart from this, the buy-to-fly ratios offered for these components at present are excessively high and result in huge material wastage.

- *Shorter production runs:* The cost of components for a mass production via conventional manufacturing reduces as the quantity of components increases. That means the smaller the quantity of components to be produced, the higher is the cost of the manufactured parts. However, AM is most suitable for manufacturing customized parts with short production runs.

- *Complex geometry:* Aerospace components are generally complex in geometry. Also, aerospace components generally require components with integrated functions. This makes CNC machining time consuming and a lot of material is wasted. AM exhibits freeform fabrication and offers high suitability for manufacturing of such components.

- *High-performance parts:* To enhance fuel efficiency and to reduce emissions, aerospace components need to be light in weight. At the same time, the safety aspects need due consideration. In this regard, aerospace components should possess high strength-to-weight ratios. Also, there is a pressing need of performance for aerospace parts, specially in extreme circumstances like extreme chemical environments, ultrahigh or ultralow temperatures. In such cases, AM can bring new opportunities to develop functional components which offer high suitability in such environments.

Many components have been successfully manufactured via AM and are in use in aerospace industry. Some of the common examples of commercial AM applications in aerospace industry are given below.

According to Pinlian Han [3], around 75% of jet engine parts can be suitably manufactured using AM techniques owing to the complex and irregular shape of the parts as depicted in Figure 16.1. With the use of AM techniques, the part performance can be enhanced owing to reduced weight density, better design, etc.

Optomec systems have produced various parts using AM for applications in aerospace, defense, electronics, etc. since 1997. For the aerospace industry, Optomec has been in operation since 2011 and has produced various complex parts for jet engines, helicopters and satellites [4]. With the use of the Aerosol Jet printer the electronic printing of 3D surfaces has become possible. Arcam utilized its EBM systems for developing various function-specific parts made of lighter metals (for example, titanium) for military aircraft, missiles and subsystems and space applications [1, 5]. Specific examples include fabrication of aerospace engines, turbine blades, structural airframe components, etc. using Q20plus printers with higher accuracy at lower cost. DMG Mori [6] utilized the LMD process and developed the LASERTEC 65 hybrid additive manufacturing (HAM) 3D printer. The HAM system utilizes the LMD for deposition and milling for subtractive operations. A turbine housing made of stainless steel is an example of component development using this HAM system.

In this way, there are several examples from the aerospace industry and also from academia for manufacturing of aerospace components using AM. The major AM processes utilized for manufacturing aerospace components include fused deposition modelling (FDM), powder bed fusion (PBF), 3-dimensional printing (3DP), stereolithography (SLA), etc.

16.3.1 Repair Work of Aerospace Components Using AM

The application of AM in aerospace is not only for manufacturing of newer parts but also in repair type of work for aircraft engine parts. The main reasons behind the repairing of damaged aerospace components via the AM route are discussed in subsequent lines. Aerospace components such as

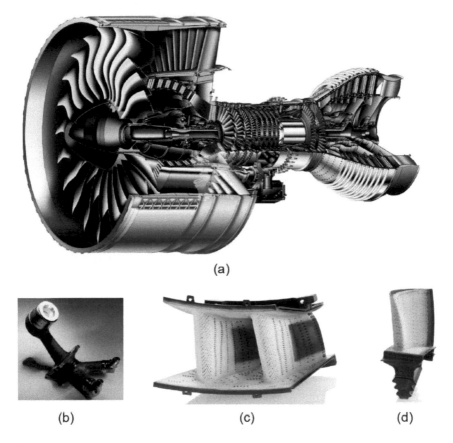

(a)

(b)   (c)   (d)

FIGURE 16.1
Application of AM in Jet Engine: (a) Turbo Fan; (b) Nozzle (Fuel); (c) Turbine Nozzle; (d) Turbine Blade. (*Source:* [3]. Han, Additive design and manufacturing of jet engine parts. *Engineering*, 2017, 3(5): pp. 648–652, under Creative Commons Licence.)

turbine blades, blisks, etc. are generally made of high performance materials and are quite expensive. Aerospace parts are generally subjected to extreme conditions leading to occurrence of wear and tear of several components. Owing to the high cost involved in fabrication, these components should be repaired to enhance their serviceability instead of their replacement. Conventionally, welding processes are generally used for such repair work. However, owing to elevated temperatures during the fusion welding processes, residual stresses are developed in the repaired part. Electron beam welding and plasma arc welding can be used to replace conventional fusion welding processes to avoid residual stress development, however, the experimental set-up required for these processes is very expensive.

AM can be successfully used to repair the damaged parts of aerospace components by utilizing hybrid AM systems. A typical HAM system and its general overview for repair work is presented in Figures 16.2 and 16.3 respectively. A detailed discussion on the principle of the working of HAM systems for repair work is explained in Chapter 15. A typical example of repair work is shown in Figure 16.4.

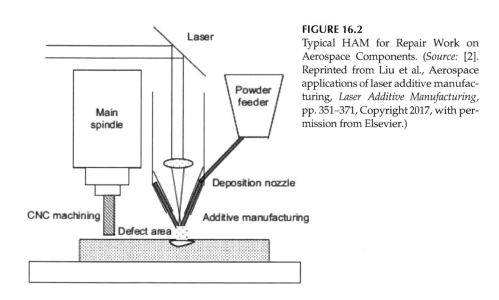

FIGURE 16.2
Typical HAM for Repair Work on Aerospace Components. (*Source:* [2]. Reprinted from Liu et al., Aerospace applications of laser additive manufacturing, *Laser Additive Manufacturing,* pp. 351–371, Copyright 2017, with permission from Elsevier.)

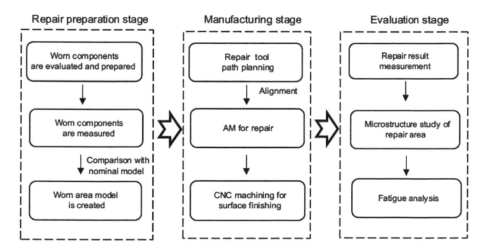

FIGURE 16.3
Overview of Repair Process Using HAM, i.e. Additive and Subtractive Operations. (*Source:* [2]. Reprinted from Liu et al., Aerospace applications of laser additive manufacturing, *Laser Additive Manufacturing,* pp. 351–371, Copyright 2017, with permission from Elsevier.)

(a)   (b)   (c)

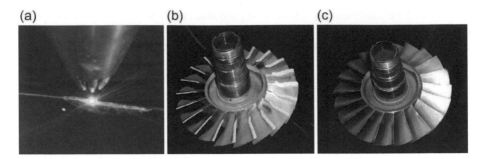

FIGURE 16.4
T700 Blisk Repair Using Laser Engineering Net-Shape Process: (a) In-Process Repair of Leading
Edge of Ti64 Airfoil; (b) Blisk after Deposition; (c) Blisk after Finishing. (*Source:* Courtesy of
Optomec Inc.)

16.4 Automotive Applications of AM

Shorter development cycle, less weight of developed product, reduced
wastage of material and considerably lower manufacturing cost involved with
AM techniques make it a suitable candidate for automotive applications. For
improving performance of existing vehicles and facilitating new vehicle
designs, a lot of experiments are carried out to justify the sustainability of processes, which is quite costly and a difficult task to be accomplished. AM offers
greater flexibility and fewer restrictions for such innovations as compared to
traditional manufacturing processes. This characteristic of AM is highly useful
for designing the vehicles with customized features. Also, AM needs less
overall lead time in absence of need for a tooling system since it directly produces final parts/shapes, which improves market responsiveness. Further, it
is now well accepted that AM reduces scrap as it utilizes only that amount of
material which is used to produce the component. All these characteristics of
AM make it suitable for its utilization in the automotive industry.

The automotive industry is one of the maiden industries for AM applications. General Motors has been utilizing AM for more than 20 years in making
prototypes for product developments. The applications of AM are not
restricted to making prototypes. It is being utilized in making various structural and functional parts such as components of gear boxes, drive shafts,
engine exhausts, etc. Kor Ecologic developed a car using AM (unveiled in
2011) named Urbee. The exterior and interior part of Urbee is completely
printed using AM. By using AM technology, the overall weight of the vehicle
is reduced due to the elimination of excess parts and joints. Fabrication of
tools and other parts is another application of AM. BMW utilizes this technology for manufacturing handheld tools used in the attachment of licence
plates and bumper [7].

A notable example of an automotive application on an SLA machine is a mammoth SLA modeller developed by Materialise possessing build volume $2100 \times 680 \times 800$ (all dimensions being in millimetres). This is sufficiently large for manufacturing most large automobile components. The outer shell of the "Areion" racing car (Formula Group T) was fabricated on this modeller in around 21 days.

In addition to commercial vehicle parts, specific vehicles such as in motorsports demand lightweight components made of advanced materials like titanium-alloys. Such components are complex in geometry and need to be produced in lower volume. Various companies have utilized AM to develop functional components for such vehicles. CRP Technology (Italy)

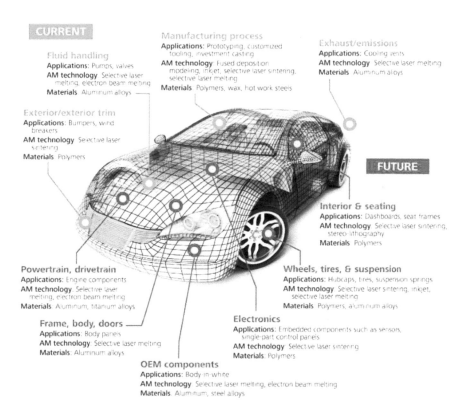

FIGURE 16.5
Current and Probable Applications of AM in Automotive Industry. (*Source:* [9, 10]. Reproduced with permission from Cotteleer et al., *The 3D opportunity primer: the basics of additive manufacturing*, 2014, Deloitte University Press, Westlake, Texas. Available from: http://dupress.com/articles/the-3d-opportunity-primer-the-basics-of-additive-manufacturing, accessed February 8, 2019. Giffi and Gangula, *3D opportunity for the automotive industry: Additive manufacturing hits the road*, 2014, Deloitte University Press, Westlake, Texas. Available from: www2.deloitte.com/insights/us/en/focus/3d-opportunity/additive-manufacturing-3d-opportunity-in-automotive.html, accessed February 8, 2019.)

has successfully utilized various AM techniques to develop components of motorsports. These parts mainly include motorbike dashboards, MotoGP engines, cam shaft covers, F1 gearboxes, suspension systems, etc. By utilizing AM techniques, significant advantages were achieved. For example, nearly 20–25% weight saving and around 25% reduction in volume was obtained via these design and fabrication techniques in the F1 gearbox [1]. Optomec developed drive shaft spiders and suspension mounting brackets (made of Ti-6Al-4V) for Red Bull Racing cars which resulted in more than 90% material reduction with considerably reduced cost and time [4]. Concept Laser designed and produced various components (made of aluminium and steel) including engine blocks, valve blocks, wheel suspensions, oil pump housings, etc. using SLM [8]. Arcam produced parts using EBM, such as gearboxes, suspension, as well as engine components possessing racing car lattices.

A (non-exhaustive) summary of AM applications, current as well as probable in the automotive sector, is given in Figure 16.5.

Despite the numerous advantages and applications of AM techniques in the automotive sector, there are some challenges which AM techniques are facing in manufacturing of automotive parts. These mainly include the limited volume of AM modellers/3D printers which restricts the production of larger components such as body panels, but significant research is in progress to overcome this problem.

16.5 Medical Applications

AM has achieved tremendous success in the medical sector over the last 20 years to the extent that this sector has become a leader in AM applications. AM is being increasingly applied for medical appliances; tissue engineering scaffold, pharmaceutical, ex-vivo tissues, medicine delivery systems, medical devices, surgical implants, prosthetics, medical training models, orthodontic and orthopaedic implants, as well as multiple medical equipments, are presently manufactured by AM processes. According to Wohlers Report in 2012, the revenue of AM in medical applications was reported as 16.4% of total revenue of the AM market [11]. The major reason behind this growth is the suitability of AM products in the medical sector. Several small but simultaneously value-dense parts like dental crowns, implants used in surgery, hearing aids, etc. can suitably be produced using AM techniques. Successful fabrication of ears, bones, muscles, etc. has been reported by various researchers so far and the same has been successfully tested on animals. Some medical systems have gained necessary clearances while others are in the development stage. The environment of AM techniques in their medical applications is such as to allow minimal wastage and easy part fabrication. Apart from the processing step, the basic AM steps are followed for these applications. The 3D CAD data is usually derived from results of MRI and CT scans of the input source. DICOM and MIMICS is then required to finally obtained .stl files and simulate them respectively in order to be successfully utilized by the AM modeller for part fabrication. The process chain to obtain the medical parts is presented in Figure 16.6. The five standard steps to be followed in testing and characterizing parts for medical applications are:

1. Convert DICOM file to 3D data
2. Obtain .stl file
3. Select appropriate AM technique
4. Select process parameters
5. Characterize and test.

Some examples of AM applications in medical sector are described here. Recently, Stratasys has started to manufacture special coils for magnetic resonance imaging (MRI) machines using the FDM technique. These coils are conventionally manufactured using injection molding and CNC machining which incurs high cost and time. The use of FDM for developing these

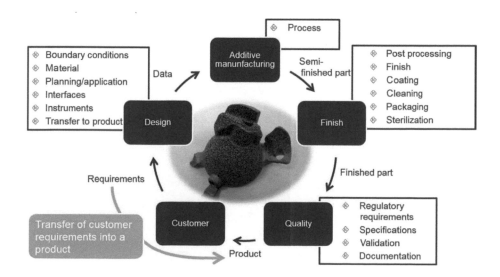

FIGURE 16.6
Process Chain for Obtaining Medical Models. (*Source:* [12]. Reprinted from Munsch, Laser additive manufacturing of customized prosthetics and implants for biomedical applications, *Laser Additive Manufacturing*, pp. 399–420, Copyright 2017, with permission from Elsevier.)

plastic components reduces the manufacturing time and cost drastically. The PolyJet technology of Stratasys is used in development of multi-color models for clinical training. AM technology enables development of 3D models of human body parts, which support the surgeon to visualize and provide great help during surgical procedures by virtue of providing previously planned surgical cuts. Arcam has launched Q10plus printers for manufacturing of orthopaedic implants.

A few applications of bio-printing, implants, scaffolds, organ visualization and prostheses are shown in Figures 16.7–16.12.

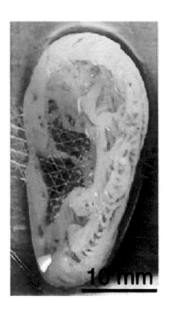

FIGURE 16.7
Bioprinted Ear Cartilage. *Source:* [13]. Reprinted from Kang et al., A 3D bioprinting system to produce human-scale tissue constructs with structural integrity. *Nature Biotechnology*, 2016, 34: p. 312, Springer, with permission from Springer Nature.

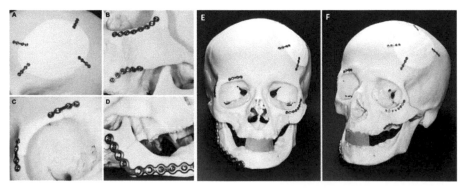

FIGURE 16.8
Examples of Various Implants Fabricated Using AM. (*Source:* [14]. Reprinted from Klammert et al., 3D powder printed calcium phosphate implants for reconstruction of cranial and maxillofacial defects. *Journal of Cranio-Maxillofacial Surgery*, 38(8): pp. 565–570, Copyright 2010, with permission from Elsevier.)

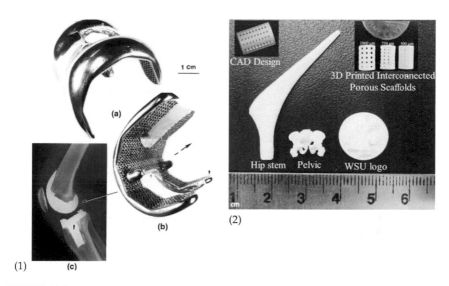

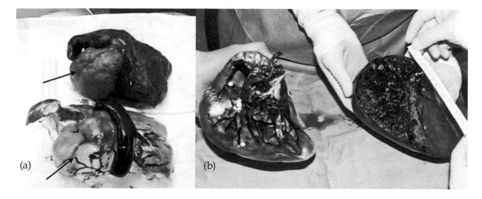

FIGURE 16.9

(1) Parts (a) and (b) Femoral (knee) Prototypes in Annealed and Polished Form Built with Porous Mesh Features, in (b) larger arrow shows the build direction; (c) A Commercially Implanted CoCrMo Femoral (Knee) Component (arrow) and a Ti–6Al–4V Tibial (Knee) Component (at t). (2) Hip Stem, Human Pelvis and WSU Logo Printed by Binder Jetting 3D Printer Using TCP (Fabricated at Washington State University). (*Source:* (1) [15]; (2) [16]. Reprinted from Gaytan et al., Comparison of Microstructures and Mechanical Properties for Solid and Mesh Cobalt-Base Alloy Prototypes Fabricated by Electron Beam Melting. *Metallurgical and Materials Transactions A,* 41(12): pp. 3216–3227, Copyright 2010, with permission from Elsevier. Bose and Tarafder, Calcium phosphate ceramic systems in growth factor and drug delivery for bone tissue engineering: a review. *Acta Biomaterialia,* 2012, 8(4): pp. 1401–1421.)

FIGURE 16.10

(a) Actual Liver of a Recipient and 3D-Printed Liver. The arrows indicate a regenerative nodule in both 3D-printed liver and native liver. (b) Pre-Operatively 3D-Printed Right Lobe and Actual Right Lobe of a Donor. (*Source:* [17]. Reprinted from Zein et al., Three-dimensional print of a liver for preoperative planning in living donor liver transplantation. *Liver Transplantation,* 2013, 19(12): pp. 1304–1310, with permission from Wiley.)

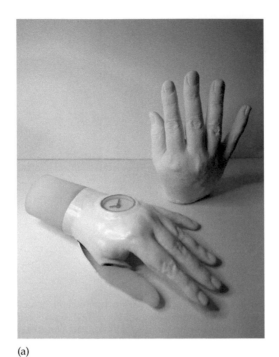

(a)

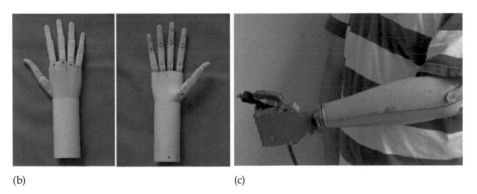

(b)   (c)

FIGURE 16.11
Examples of 3D Printed Upper Limb Prostheses: (a) Scand: Passive Adjustable Forearm Prosthesis; (b) Prosthetic Hand Prototype: An Externally Powered Forearm Prosthesis; (c) Gosselin's Hand: A Body Powered Forearm Prosthesis. (*Source:* (a) [18]. Reproduced with permission from Scand, 3D Printing Prosthetic Hand – Make It Real Challenge. Available from: www.instructables.com/id/3D-Printing-Prosthetic-Hand-Make-it-Real-Challen, accessed on January 19, 2019; (b) [19]. Andrianesis and Tzes, Development and control of a multifunctional prosthetic hand with shape memory alloy actuators. *Journal of Intelligent and Robotic Systems*, 2015, 78(2): pp. 257–289; (c) [20]. Laliberté et al., Towards the design of a prosthetic underactuated hand. *Mechanical Sciences*, 2010, 1(1): pp. 19–26).

FIGURE 16.12
Example of Prosthesis Developed via AM: (a) in Open Position; (b) in Closed Position Holding a Pen; and (c) in Closed Position Holding a Bottle. (*Source:* [21]. Reproduced with permission from Ferreira et al., Ferreira, I., Development of low-cost customised hand prostheses by additive manufacturing. *Plastics, Rubber and Composites*, 2018, 47(1): pp. 25–34.)

16.5.1 Applications of AM in Biomaterials

If the advantages of AM techniques are combined with the utilization of biomaterials, many promising person-specific applications in the medical field can be facilitated. Biomaterials are an exquisite class of materials that can restore functions of human tissues and can be either natural or man-made. There are a variety of materials like aluminium oxide for dental implants, nickel titanium alloys for catheters and many others which have been cleared by the FDA for various medical applications.

Thus, from the above discussion, it is clear that AM has huge applications and scope in the medical sector. The common criteria for which AM applications suit the medical sector and a summary of major areas of AM applications are presented in Tables 16.1 and 16.2 respectively.

TABLE 16.1

Criteria that Favours AM Applications in Medical Sector

Sr. no.	Criteria	Achievements
1	Designing and manufacturing of surgical aid tools, bio-models and implants	• Play a useful role for design and production of surgical support tools, bio-models and implants • Also used to upgrade surgical tools
2	Designing and manufacturing of various scaffolds for tissue engineering	• An AM characteristic is to design and manufacture the different scaffolds for restoration of tissues • It replaces conventional scaffold fabrication methods • AM help for printing organs, produced cells, cell-laden biomaterials, biomaterials individually
3	Development of various medical devices and surgical training models	• Used for developing various medical models, surgical training models which are used in medical education
4	Individualization	• AM is used for individualization, as data differ from patient to patient; used for customized implant fabrication
5	Complex geometries	• This technology has great potential for fabrication of complex geometries implant
6	Functional integration	• The medical models are in functional integration • It works like an original one
7	Reduced costs	• AM technology helps to reduce cost of medical implant as compared to other manufacturing processes
8	Rapid availability	• AM has availability for producing medical model in short time
9	Improved patient care	• This technology is used for development and improvement of patient care through the customized model
10	Cost-effectiveness for the hospital	• AM technology makes possible manufacturing of customized implants which comfortably fit the patient at reasonable cost
11	Weight reduction	• Reduction of weight done with the help of this technology by changing material

Source: [22]. Javaid and Haleem, Additive manufacturing applications in medical cases: a literature based review. *Alexandria Journal of Medicine*, 2018, 54(4): pp. 411–422.

16.6 Construction Industry

Utilization of AM to print houses and villas has recently been started, but this application has still not fully matured. Continuous research is being carried out in the construction industry to explore AM to the fullest extent and more than 30 research groups are currently engaged in various related R&D activities [23]. Figure 16.13 shows the rise of AM application in the construction industry since the seminal work of Pegna in 1997 [24]. The main AM techniques suitable for construction industry include SLA, FDM, IJP, SLS and contour crafting. Some novel 3D printing processes have also been developed recently. For example, at Loughborough University, a concrete printing process has been developed which utilizes a similar principle of extrusion of cement mortar as that used by contour crafting.

In the construction industry, AM offers several benefits which mainly include [25]:

- *Reduced manpower:* Owing to the highly automated process, the manpower requirement is relatively less.
- *Design flexibility:* High degree of design flexibility is offered as compared to manual construction practice.

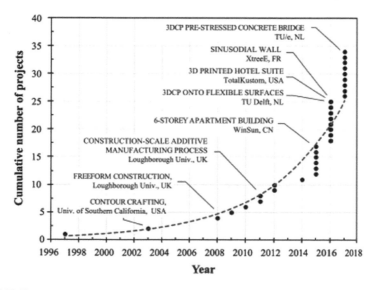

FIGURE 16.13
Development and Rise of AM Applications in Construction Industry since Concept Inception in 1997. (*Source:* [23]. Buswell et al., 3D printing using concrete extrusion: a roadmap for research. *Cement and Concrete Research*, 2018, 112: pp. 37–49, under Creative Commons Licence.)

TABLE 16.2

Major areas of AM Applications in Medical Sector

Sr. no.	Area of medical application	Objectives	Major benefits
1	Surgical planning	• Main objective is how AM become more beneficial in surgical planning • These models provide surgical and physician team with visual aid used to improve surgery planning • Bone structure of patient is studied before surgery, which reduces operation time, cost, as well as risk	• With the help of this technology during operation, the problem cause was predicted and diagnostic quality obtained • AM models are better to understand the complex anomaly and complicated procedure • These models especially help in surgeries where there are deformities or anatomical abnormalities, and in heart, spine, maxillofacial and craniofacial surgery
2	Medical education and training	• The primary purpose is that this technology provides better demonstration of internal and external human anatomy structure • It consists of many colors, so these models are used in teaching as well as research in medical education	• AM models used for better illustration in school and museums • These models are used by young doctors or medical students to understand surgical procedures and problems without causing patient discomfort
3	Design and development of devices and instrumentation used in medicine	• The purpose of this technology is how this helps for design and development of devices and instrumentation used in medicine	• For fabrication of medical devices and instrumentation AM is used because this technique designs the model, develops and then produces required medical equipment or instruments • It includes hearing aids, dental devices and surgical tools
4	Customized implant design	• The purpose of this technology has potential to fabricate customized fixtures and implants • Complex geometry is also built in short time	• CAD and AM technology make possible manufacture of customized implants which comfortably fit the patient with reasonable cost • AM creates accurate implants for patient rather than standard-sized implants such as knee joints, spinal implants and dental implants which is significantly beneficial for patient. Surgical implants become more precise by using AM • With customized implant fabrication risk and surgery time is reduced

5	Scaffoldings and tissue engineering	• The primary purpose is how this technology fabricates implants with its unique geometrical characteristics like scaffolds for the restoration of tissues • It replaces conventional scaffold fabrication methods	• The scaffold is supporting structure and provides support and guidance to defective patient bone or growing tissue which is damaged • AM techniques like FDM, SLS and 3D printing are suitable for fabricating controlled porous structures by using application of biomaterial contributing to the field of tissue engineering and scaffolding • AM technology increases the ability to produce complex geometry product with higher accuracy
6	Prosthetics and orthotics	• How this technology is beneficial in prosthetics and orthotics field of medical which starts with particular patient anatomy	• Accurate alignment characteristics of patient also needed in this model, allowing biomechanically correct geometry development and improving comfort, stability • AM fabricates custom prosthesis which fit precisely to patient at reasonable cost, such as pattern of dental crowns
7	Mechanical bone replicas	• How AM technology used for mechanical bone model fabrication • This technology easily replicates material variation	• SLA can create composite structure which has similar property to bone • These bones can be provided with strength under various conditions • Also beneficial to recreate the stresses, fractures and different changes in bone, which gives more help to researchers and doctors
8	Forensics	• AM tool is more beneficial tool for investigation of criminals, such as homicide cases where crime scene for investigation is reconstructed	• These models aid in keeping evidence related to criminal investigations. By analyzing scaffolds, the investigator can solve a lot of underlying questions

Source: [22]. Javaid and Haleem, Additive manufacturing applications in medical cases: a literature based review. *Alexandria Journal of Medicine*, 2018, 54(4): pp. 411–422.

- *Reduced waste:* Considerably less scrap or waste material.
- *Construction time:* Significantly less construction time as compared to conventional construction practices.

Owing to limitations upon the build size of AM modellers, a general conclusion was that medium-sized and large buildings are difficult to develop using these techniques. However, recent research/project work shows the developments in large-scale 3D printers that can enable the printing of full buildings. Three major projects showing such developments are discussed here. In 2014, WinSun (a Chinese architectural company) developed house groups of $200\,m^2$ each using AM in a day. Modeller dimensions utilized in printing this building was $150 \times 10 \times 6.6$ meters (length $\times$ width $\times$ height). In 2014, Qindao Unique Technology revealed a huge AM modeller of dimensions $12 \times 12 \times 12$ meters; this worked on the principle of FDM technology that deposits layers of semi-molten material which were subsequently bond together by a process similar to diffusion bonding. Similarly, in 2015, a villa as well as one five-storey apartment could successfully be developed via the AM route. In these constructions, printing of individual building components was accomplished at different locations. Individual parts were finally brought to the site to suitably assemble and install them to obtain the final building structure. This project was an excellent demonstration of AM application to print a complete building [25]. Several such projects have been completed and various others are under development. A brief summary of AM development in the construction sector is presented in Table 16.3. An example of 3D structures obtained via the AM route are shown in Figure 4.2.

TABLE 16.3

Recent Developments in 3D printing in Construction Industry

Studies	3-D printing technology presented	Printed products
Hinczewski et al. [26]	Stereolithography	Ceramic part
Khoshnevis et al. [27]	Contour crafting	Plaster part Ceramic part
Ryder et al. [28]	Concept modelling	Polyester part
Lim et al. [29]	Concrete printing	Concrete part
Gibson et al. [30]	FDM and SLS	Space frame architectural model Rotunda architectural model IBM Pavilion architectural model
Dimitrov et al. [31]	3DP	Plaster model
Bogue [7]	KamerMaker	Large-scale polypropylene components
Kietzmann et al. [32]	3DP	Entire house
Liang et al. [33]	FDM	Entire house

Source: [25]. Wu et al., A critical review of the use of 3-D printing in the construction industry. *Automation in Construction*, 2016, 68: p. 21–31.

16.7 Retail Applications

In the retail market, 3D printed objects such as clothing, shoes and other consumer goods have already captured the attraction of end users. AM offers great freedom for design and creation of complex and intricate designs for the fashion industry. With the application of 3D printing, it is expected that the retail industry will certainly benefit. In the opinion of John Hauer (co-founder and CEO of 3-DLT) [34], products in the retail market will reach market swiftly with better design through the use of AM owing to the reduction in supply chain costs and freedom of innovation, as well as visualization of product models. Different types of designs have a high cost to the manufacturer for their development using conventional manufacturing techniques. AM enables designs to be prototyped in real time even in small quantities. Once the prototype is made, then mass production can be carried out using conventional manufacturing processes.

AM is being used in the design and development of fashion products such as jewellery, apparels, etc. Continuum (a retailer) offers AM products like shoes, jewellery, bathrobes, etc. which are customized. The commonly used AM processes for retail market include SLS, FDM, binder jetting, polyjet printing and stereolithography [35]. Table 16.4 presents a brief summary of these commonly used AM techniques in the fashion industry, along with benefits, challenges, commonly used materials and main brands utilizing these products. However, although AM offers freedom to designers to develop prototypes for customized designs at reduced lead time, there are several challenges to be addressed before its complete utilization in retailing.

16.8 Summary

A summary of various applications and benefits gained by various industries upon utilization of various AM techniques is presented in Table 16.5.

In this chapter, the domain of AM applications in various relevant sectors like aerospace, automotive, medical, biomaterials, food, retail, fashion industries, etc. have been discussed along with suitable examples. The next chapter provides readers with a detailed discussion on the impact and forecasting of additive manufacturing.

TABLE 16.4

Comparison of Five AM Processes Utilized in Fashion Industry

AM method	Printer company	Maximum printing size	Materials	Benefits	Challenges	Product categories	Brands or designers
SL	Mammoth Materialise	2 m	Photopolymer resins	Large objects; detailed objects; fast lead time; user-friendly; high-quality surface finish	Support rafts; large space needed	Long dresses; detailed components	Iris van Herpen; Lady Gaga
SLS	PRECIOUS M 080 EOS	80 mm × 95 mm	Metal powder	No support rafts; compact size; fast lead time; high-quality surface finish	Limited printing size; limited end-uses	Jewellery; watches; metal accessories	Dr Richard Hoptroff
FDM	Objet Connex Stratasys	490 mm × 390 mm × 200 mm	Liquid wax, metal, and ceramic filaments	Multi-material; compact size; user-friendly; detailed objects; multiple products at once	Support rafts; lower surface quality	Highly textured dresses and separates	Iris van Herpen
FDM	Replicator Desktop MakerBot	200 mm × 250 mm × 150 mm	PLA filament; ABS filament	Flexible materials; compact size; user-friendly; multiple products at once; various quality levels	Support rafts; slower lead time; limited printing size	Dresses; accessories; garment components; prototypes	Francis Bitonti
3DP	Spectrum Z510 3D Systems	254 mm × 356 mm × 203 mm	Powdered metal and ceramic filaments	Inexpensive; prints in color; fastest lead time; high-quality surface finish	Weaker products	Shoes; accessories; prototypes	Timberland

Source: [35]. Vanderploeg et al., The application of 3D printing technology in the fashion industry. International Journal of Fashion Design, *Technology and Education,* 2017. 10(2): p. 170–179.

TABLE 16.5

Industries Benefiting from AM

Industry	Applications	Benefits gained
Aerospace	• Prototyping • Component manufacturing • Reducing aircraft weight • Engine components for the Airbus • Flight-certified hardware • Manufacturing of satellite components	• Produce very complex work pieces at low cost • Allow product lifecycle leverage • Objects manufactured in remote locations, as delivery of goods is no longer a restriction • Reduction in lead-time would imply reduction in inventory and reduction in costs • On-demand manufacturing for astronauts • Eliminate excess parts that cause drag and add weight • Improve quality
Automotive	• Prototyping • Component manufacturing • Reducing vehicle weight • Cooling system for racing car	• Help eliminate excess parts • Speed up time to market • Reduce cost involved in product development • Reduce repair costs considerably • Reduce inventory • Could effectively change the way cars will look and function in future • Improve quality
Machine tool production	• Prototyping • Reducing grip system weight • End-of-arm for smarter packaging	• Quick production of exact and customized replacement parts on site • Allow for designs that are more efficient and lighter
Healthcare and medical	• Fabricating custom implants, such as hearing aids and prosthetics • Manufacturing human organs • Reconstructing bones, body parts • Hip joints and skull implants • Robotic hand	• Reduced surgery time and cost • Reduced risk of post-operative complications • Reduced lead-time

Continued

TABLE 16.5 *Continued*

Industry	Applications	Benefits gained
Dentistry and dental technology	• Dental coping • Precisely tailored teeth and dental crowns • Dental and orthodontic appliances • Prototyping	• Great potential in use of new materials • Reduced lead-time • Prosthetics could be fabricated in only a day, sometimes even in a few hours
Architectural and construction	• Generating exact scale model of the building • Printing housing components	• Producing scale models up to 60% lighter • Reduce lead-times of production by 50—80% • Ability to review a model saves valuable time and money caused by reworking • Reduce construction time and manpower • Increase customization • Reduce construction cost to provide low cost housing to poverty-stricken areas
Retail/apparel	• Shoes and clothing • Fashion and consumer goods • Consumer grade eyewear • Titanium eyeglass frames • Production of durable plastic and metal bicycle accessories	• On-demand custom fit and styling • Reduce supply chain costs • Create and deliver products in small quantities in real time • Create overall better products • Products get to market quicker
Food	• Chocolate and candy • Flat foods such as crackers, pasta and pizza	• Ability to squeeze out food, layer by layer, into 3-D objects • Reduce cost • Feasibility of printing food in space

Source: [36]. Attaran, M., The rise of 3-D printing: the advantages of additive manufacturing over traditional manufacturing. *Business Horizons*, 2017, 60(5): pp. 677–688.

References

1. Guo, N., Leu, M. C., Additive manufacturing: technology, applications and research needs. *Frontiers of Mechanical Engineering*, 2013, 8(3): pp. 215–243.
2. Liu, R., Wang, Z., Sparks, T., Liou, F., Newkirk, J., Aerospace applications of laser additive manufacturing, in *Laser additive manufacturing*, M. Brandt, Editor, 2017, Woodhead Publishing, pp. 351–371.
3. Han, P., Additive design and manufacturing of jet engine parts. *Engineering*, 2017, 3(5): pp. 648–652.
4. Optomec, http://www.optomec.com/optomec-overview/, accessed on February 5, 2019.
5. Arcam A B., http://www.arcam.com, accessed on February 10, 2019.
6. Mori, D., *Additive manufacturing for quality finished parts*, 2014. Available from: http://us.dmgmori.com/blob/354972/7009c74e7ad96fad7ad34555d8 1bc316/pl0us14-lasertec-65-3d-pdf-data.pdf.
7. Bogue, R., 3D printing: the dawn of a new era in manufacturing? *Assembly Automation*, 2013, 33(4): pp. 307–311.
8. www.concept-laser.de.
9. Cotteleer, M., Holdowsky, J., Mahto, M., *The 3D opportunity primer: the basics of additive manufacturing*, 2014, Deloitte University Press, Westlake, Texas. Available from: http://dupress.com/articles/the-3d-opportunity-primer-the-basics-of-additive-manufacturing, accessed February 8, 2019.
10. Giffi, C. A., Gangula, B., *3D opportunity for the automotive industry: Additive manufacturing hits the road*, 2014, Deloitte University Press, Westlake, Texas. Available from: www2.deloitte.com/insights/us/en/focus/3d-opportunity/additive-manufacturing-3d-opportunity-in-automotive.html, accessed February 8, 2019.
11. Wohlers, *Additive manufacturing and 3D printing: state of the industry*, 2013.
12. Munsch, M., Laser additive manufacturing of customized prosthetics and implants for biomedical applications, in *Laser additive manufacturing*, M. Brandt, Editor, 2017, Woodhead Publishing, pp. 399–420.
13. Kang, H.-W., Lee, S. J., Ko, I. K., Kengla, C., Yoo, J. J., Atala, A., A 3D bioprinting system to produce human-scale tissue constructs with structural integrity. *Nature Biotechnology*, 2016, 34: p. 312.
14. Klammert, U., Gbureck, U., Vorndran, E., Rödiger, J., Meyer-Marcotty, P., Kübler, A. C., 3D powder printed calcium phosphate implants for reconstruction of cranial and maxillofacial defects. *Journal of Cranio-Maxillofacial Surgery*, 2010, 38(8): pp. 565–570.
15. Gaytan, S. M., Murr, L. E., Martinez, E., Martinez, J. L., Machado, B. I., Ramirez, D. A., Medina, F., Collins, S., Wicker, R. B., Comparison of microstructures and mechanical properties for solid and mesh cobalt-base alloy prototypes fabricated by electron beam melting. *Metallurgical and Materials Transactions A*, 2010, 41(12): pp. 3216–3227.
16. Bose, S., Tarafder, S., Calcium phosphate ceramic systems in growth factor and drug delivery for bone tissue engineering: a review. *Acta Biomaterialia*, 2012, 8(4): pp. 1401–1421.
17. Zein, N. N., Hanouneh, I. A., Bishop, P. D., Samaan, M., Eghtesad, B., Quintini, C., Miller, C., Yerian, L., Klatte, R., Three-dimensional print of a liver for preoperative planning in living donor liver transplantation. *Liver Transplantation*, 2013, 19(12): pp. 1304–1310.

18. Scand, A. S., 3D Printing Prosthetic Hand – Make It Real Challenge. Available from: www.instructables.com/id/3D-Printing-Prosthetic-Hand-Make-it-Real-Challen, accessed on January 19, 2019.

19. Andrianesis, K., Tzes, A., Development and control of a multifunctional prosthetic hand with shape memory alloy actuators. *Journal of Intelligent and Robotic Systems*, 2015, **78**(2): pp. 257–289.

20. Laliberté, T., Baril, M., Guay, F., Gosselin, C., Towards the design of a prosthetic underactuated hand. *Mechanical Sciences*, 2010, **1**(1): pp. 19–26.

21. Ferreira, D., Duarte, T., Alves, J. L., Ferreira, I., Development of low-cost customised hand prostheses by additive manufacturing. *Plastics, Rubber and Composites*, 2018, **47**(1): pp. 25–34.

22. Javaid, M., Haleem, A., Additive manufacturing applications in medical cases: a literature based review. *Alexandria Journal of Medicine*, 2018, **54**(4): pp. 411–422.

23. Buswell, R. A., Leal de Silva, W. R., Jones, S. Z., Dirrenberger, J., 3D printing using concrete extrusion: a roadmap for research. *Cement and Concrete Research*, 2018, **112**: pp. 37–49.

24. Pegna, J., Exploratory investigation of solid freeform construction. *Automation in Construction*, 1997, **5**(5): pp. 427–437.

25. Wu, P., Wang, J., Wang, X., A critical review of the use of 3-D printing in the construction industry. *Automation in Construction*, 2016, **68**: pp. 21–31.

26. Hinczewski, C., Corbel, S., Chartier, T., Stereolithography for the fabrication of ceramic three-dimensional parts. *Rapid Prototyping Journal*, 1998, **4**(3): pp. 104–111.

27. Khoshnevis, B., Russell, R., Kwon, H., Bukkapatnam S., Crafting large prototypes. *IEEE Robotics and Automation Magazine*, 2001, **8**(3): pp. 33–42.

28. Ryder, G., Ion, B., Green, G., Harrison, D., Wood, B., Rapid design and manufacture tools in architecture. *Automation in Construction*, 2001, **11**(3): pp. 279–290.

29. Lim, S., Buswell, R. A., Le, T. T., Austin, S. A., Gibb, A. G. F., Thorpe, T., Developments in construction-scale additive manufacturing processes. *Automation in Construction*, 2012, **21**: pp. 262–268.

30. Gibson, I., Kvan, T., Ming, L. W., Rapid prototyping for architectural models. *Rapid Prototyping Journal*, 2002, **8**(2): pp. 91–95.

31. Dimitrov, D., Schreve, K., de Beer, N., Advances in three dimensional printing: state of the art and future perspectives. *Rapid Prototyping Journal*, 2006, **12**(3): pp. 136–147.

32. Kietzmann, J., Pitt, L., Berthon, P., Disruptions, decisions, and destinations: enter the age of 3-D printing and additive manufacturing. *Business Horizons*, 2015, **58**(2): pp. 209–215.

33. Liang, F. F., Liang, R. W., Xiang, X. H., Analysis of effect factors about the ground settlement during foundation pit excavation. *Advanced Materials Research*, 2015, **1065–1069**: pp. 7–10.

34. Honigman, B., How 3D printing is reinventing retail. *Forbes Magazine*, 2014. Available from: https://www.forbes.com/sites/centurylink/2014/06/03/how-3d-printing-is-reinventing-retail/#48e07123795d, accessed on August 27, 2019.

35. Vanderploeg, A., Lee, S-E., Mamp, M., The application of 3D printing technology in the fashion industry. *International Journal of Fashion Design, Technology and Education*, 2017, **10**(2): pp. 170–179.

36. Attaran, M., The rise of 3-D printing: the advantages of additive manufacturing over traditional manufacturing. *Business Horizons*, 2017, **60**(5): pp. 677–688.

17

Impact and Forecasting of Additive Manufacturing

17.1 Introduction

In the current manufacturing era, there is hardly any aspect of customer life that is untouched by the applications of AM technology. Though its initial use was restricted to prototype production, today AM has emerged as a full-fledged technology and has a broad spectrum of applications in direct and indirect prototyping, as well as tooling and manufacturing. It is gaining importance in various sectors of business as well as in the daily life of consumers.

With the growing interest and industrial application of AM, the next open question is about the comparison of AM processes with traditional manufacturing techniques, the sustainability of AM processes and their impact on different sectors of society. The detailed comparison between AM and conventional manufacturing processes has already been provided in Chapter 2. This chapter is the penultimate of Section D, which covers the trends, advancements, applications and conclusion. Various applications of AM technology in different application areas has been discussed in the previous chapter. This chapter presents a detailed discussion of the impact and forecasting of additive manufacturing, including the influence of AM upon health and well-being, environment, supply chain management, health and occupational hazards, repair; economic characteristics of AM; sustainability of AM; and the future of AM. It concludes the discussion with a summary.

17.2 Impact of AM

AM continues to transform the manufacturing paradigm, including design, as well as the way in which production capital is distributed. The AM industry is extending itself to serve society via 3D printers in homes, offices, schools, factories, etc. It will totally change the market scenario. However, the AM industry is still not established enough to troubleshoot probable hazardous health and environmental effects. This necessitates development of proactive assessment tools to enable personnel involved in material development, designing, printing operations, as well as end customers, to create as well as judge appropriateness of materials and processes that suit their specific requirements. Currently, life cycle assessment is not sufficiently competent to provide sufficient information related to raw material hazards.

The impact of different AM techniques varies according to feedstock material, particular AM techniques and post treatment processes. A summative illustration of the benefits and weaknesses of AM over conventional techniques in terms of environmental, economic and socio dimensions is presented in Figure 17.1.

17.2.1 Impact on Health and Well-Being

Since World War II, the people's general quality of life has been increased owing to easy availability of medicines, vaccines, advancements in surgical procedures, etc. The mortality rate has been reduced globally which has led to steady increase in overall population as also the increased aged population. According to NIA (National Institute on Aging, United States), around 500 million people were of 65 years of age and above in 2006; this will increase to 1 billion (projected) people by 2030 [2]. This will put a significant strain on the world-wide government budget to take care of the aging population. Under these circumstances, providing efficient and high-quality healthcare has become one of the key challenges towards improving the health as well as overall well-being of the global population [3].

A promising approach to deliver good healthcare is to offer people personalized care depending on their individual needs. AM specializes in development of customized facilities as per need of individual patients. AM suitably fabricates implants used in surgery (such as for skull, elbow, knee joint, hip joint, dentistry, etc.), medical equipment, biomaterials, etc. Detailed examples of medical and biomaterial applications of AM are presented in Chapter 16. Thus, AM plays a key position in catering to personalized healthcare needs.

17.2.2 Impact on Environment

It is well established that AM offers various environmental benefits like energy efficiency, lower material wastage, and reduction in transport impact owing to capability of on-order or on-site manufacturing as compared to traditional manufacturing. However, one of the similarities between AM and traditional manufacturing techniques is consumption of energy and material. Owing to the consumption of energy and material, emission generation during AM leaves its own environmental footprint.

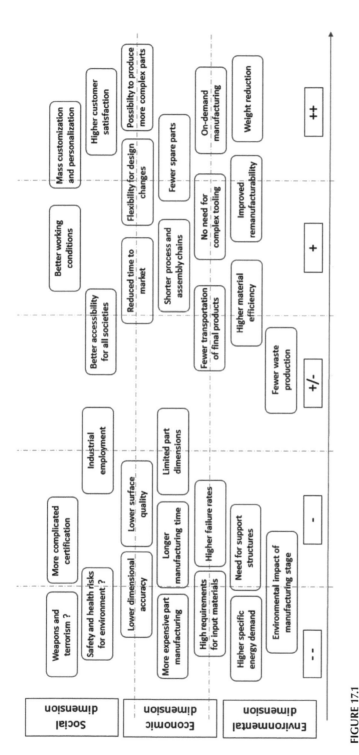

FIGURE 17.1
Benefits and Weaknesses of AM as Compared to Traditional Manufacturing Techniques. Note: + denotes benefits; – denotes weakness. (*Source:* [1]. Reproduced from Kellens et al. Environmental dimensions of additive manufacturing: mapping application domains and their environmental implications. *Journal of Industrial Ecology*, 2017, 21(S1): pp. S49–S68., under Creative Commons Licence.)

The amount of emissions largely depends on numerous factors, such as form of feedstock (powder, liquid and sheet), materials (polymers, ceramics, metals, composites, etc.), processes (from DED to PBF to material extrusion), etc.

The National Science Foundation organized a workshop in 2014 to identify the environmental implications of AM and reported five key environmental challenges of AM (see Figure 17.2) [4].

AM technology has several benefits over traditional manufacturing processes as discussed in Chapter 2.

17.2.3 AM Impact upon Supply Chain Management

The method of manufacture and distribution of products to end users during conventional manufacturing processes involve the combined efforts of various companies such as suppliers of raw material and components, manufacturer of equipments, distributors, retailers and so on. This combination of different companies constitutes the manufacturing supply chain system. In a manufacturing supply chain, products start their journey from suppliers and reach the hands of customers through various stages. AM has the capability of reducing several stages (such as warehousing, transport-

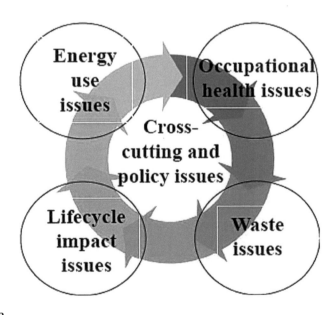

FIGURE 17.2
Key Issues of Environmental Implications of AM. (*Source:* [5]. Reprinted from Rejeski et al., Research needs and recommendations on environmental implications of additive manufacturing. *Additive Manufacturing*, Vol. 19, pp. 21–28, Copyright 2018, with permission from Elsevier.)

ation, etc.) which are generally needed in conventional manufacturing. In this direction, AM can offer freedom to redesign for a limited number of products, permits the ability to fabricate near net shaped products near customer locations, and also saves time and cost.

17.2.4 Health and Occupational Hazards

One point of similarity of AM and conventional manufacturing techniques is that both consume material as well as energy. Also, emission generation is common in both cases. These emissions are hazardous to a large extent to health as well as occupation, depending upon their intensity as discussed here. Traditional manufacturing techniques like casting, machining, forging, welding, etc. result in numerous kinds of water/air emissions, fluid spills, noise, wasted chips and so on which have potential health hazards [3]. One of the prime factors is oil mist which is a resultant of metal working fluids. Long exposure to different kinds of oil mist can cause several kinds of cancer and other diseases. Noise is also a common hazard which can result in occupational hearing loss. Thus, there are several such kinds of hazards which can come from the use of conventional manufacturing techniques. These kinds of hazards can be eliminated with the use of AM techniques. AM, on the other hand, leads to newer health issues.

A broad range of materials such as epoxy resins, polycarbonates, nylons, elastomers and so on have been processed via AM. However, the effects of some of these have not been well tested. Therefore, there is a strong need for standardization of several unexplored research areas of AM and its environmental implications, potential toxicity and chemical degradability of solvents [6]. Table 17.1 shows some of the chemicals used in AM processes and their health and occupational hazards.

17.2.5 Impact on Repair

In the future, AM will establish a significant place in repair and construction types of work. During the service of engineering parts, these parts suffer from wear and tear. It is often seen that some areas of these parts deteriorate more as compared to other areas. In that condition, it is more economic to repair/replace that area of higher deterioration than replace the complete part. AM can be utilized to selectively repair specific portions of engineering parts. The major benefit of AM technology is its multi-axial movement (up to six axes of rotation/movement). Several repair works have been successfully performed using AM techniques. These mainly include propellers and landing gears of helicopters, turbine blades, engine blocks, pistons, camshafts, numerous military devices and so on.

TABLE 17.1

Different Chemicals Utilized in AM Techniques and Their Environmental Effects

AM process	Chemical/solvent	Emissions	Hazards of usage	Biodegradability
SLA	Propylene carbonate	CO_2, CO, SOx	Low system toxicity was found in rats	Readily biodegradable (more than 80% degraded in 10 days)
	Urethane resins		Too much ingestion may lead to vomiting	Not found to be dangerous to the environment
	Tripropylene glycol		Slight irritation after eye contact No absorption or irritation to the skin	Can be biodegraded by 50% in just 8.7 days, and by 81.9% over 28-day test period
	Isopropanol		Irritation and burning sensation in eyes and sometimes corneal injuries; irritation and soreness on the skin and prolonged exposure may cause dermatitis	Has a potential to acutely reduce oxygen from aqueous systems
SLS	Polyamide resin	CO_2	No serious hazards were found during handling or exposure to this chemical	Forms inflammable mixture with some chemicals or long exposure to air
	Acrylonitrile butadiene styrene		Molten plastic likely to cause lethal burns, processing fumes may lead to eye irritation and choking of the respiratory tract	Since it is insoluble in water, its eco-toxicity is low
LENS	Photopolymers	CO_2, CO, SOx	Inhalation may cause ulcers and burning in throat and coughing; contact with skin and eyes causes redness, irritation and swelling	No hazardous decomposition products
FDM	Propylene glycol monomethylether	CO_2, CO, SOx, PMc, NOx	Irritation in eyes, skin, nose, throat; headache, nausea, dizziness, drowsiness, incoordination; vomiting, diarrhea	No hazardous decomposition products

Source: [3]. Han, Additive design and manufacturing of jet engine parts. *Engineering*, 2017, 3(5): pp. 648–652, with permission.

17.3 Economic Characteristics of AM

AM technology started its journey with rapid prototyping and has now reached a state where it allows individuals to design and develop customized products without incurring any cost of tools and molds [7]. It is now well established that technological advances affect market and firm structure.

As per Weller et al. [7], particularly for a monopoly, increasing profit margins by catering to customized product requirement by capturing end users surplus is possible by adopting AM. At the same time, owing to cut-throat market competition, firms may lower cost barrier to gain market access. In this process, they also tend to be capable of multiplexing and serving multiple markets in one go. The overall impact of both these phenomena would in general lead to lowering of prices for customers.

While AM offers a number of economic opportunities, manufacturers may face various limitations as discussed in Table 17.2.

17.4 Sustainability of AM

AM has emerged as a prominent manufacturing strategy and established itself as a key technology for varied applications in almost all engineering

TABLE 17.2

Opportunities and Drawbacks of AM from an Economic Perspective

Opportunities	Limitations
+ Acceleration and simplification of product innovation: iterations are not costly and end products are rapidly available	− High marginal cost of production (raw material costs and energy intensity)
+ Price premiums can be achieved through customization or functional improvement (e.g. light weight) of products	− No economies of scale
+ Customer co-design of products without incurring cost penalty in manufacturing	− Missing quality standards
+ Resolving "scale-scope dilemma": no cost penalties in manufacturing for higher product variety	− Product offering limited to technological feasibility (solution space, reproducibility, quality, speed)
+ Inventories can become obsolete when supported by make-to-order processes	
+ Reduction of assembly work with one-step production of functional products	− Intellectual property rights and warranty related limitations
+ Lowering barriers to market entry	− Training efforts required
+ Local production enabled	− Skilled labor and strong experience needed
+ Cost advantages of low-wage countries might diminish in the long run	

Source: [7]. Bogue, 3D printing: the dawn of a new era in manufacturing? *Assembly Automation*, 2013, 33(4): pp. 307–311.

sectors. Starting from manual craft manufacturing at the time of its inception, today it has reached industrial revolution 5.0 in the twenty-first century. AM is the most prominent manufacturing paradigm of the twentieth century. Figure 17.3 illustrates the various key events in the AM domain to highlight aspects of its sustainability timeline. Figure 17.4 presents various aspects of sustainability of AM as compared to conventional manufacturing processes.

17.5 Summary and Future of AM

AM technology is over 30 years old and has been continuously growing since its inception. However, this growth has burgeoned after the expiration of the initial patents in 2009. Several advancements have been added to this technology during these years. However, experts believe that AM has not fully matured and requires many improvements, such as reduction in the cost of 3-D printing setups, as well as range and availability of printing materials, improving the printing capabilities such as production rate, speed, accuracy, etc.

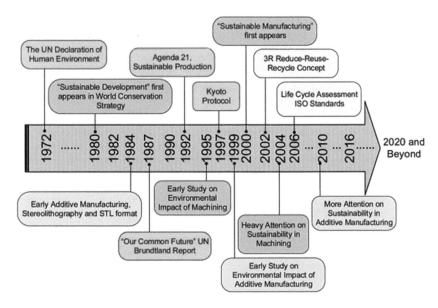

FIGURE 17.3

Illustration of Sustainability Timeline of AM Systems. (*Source:* [8]. Reprinted from Peng et al., Sustainability of additive manufacturing: An overview on its energy demand and environmental impact. *Additive Manufacturing*, Vol. 21, pp. 694–704, Copyright 2018, with permission from Elsevier.)

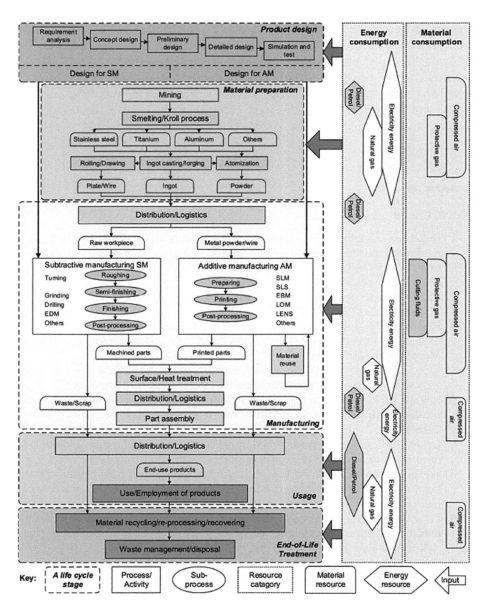

FIGURE 17.4
Sustainability of AM as Compared to Conventional Manufacturing. (*Source:* [8]. Reprinted from Peng et al., Sustainability of additive manufacturing: An overview on its energy demand and environmental impact. *Additive Manufacturing*, Vol. 21, pp. 694–704, Copyright 2018, with permission from Elsevier.)

Despite such a tremendous growth, advancements and benefits of AM experts do not believe that AM technology will replace traditional manufacturing processes. Instead, AM technology will complement conventional manufacturing. AM has presented newer avenues and opportunities for manufacturing, as well as globally managing supply chains.

This chapter provides readers with a detailed discussion of the impact and forecasting of AM. The various aspects of AM presented through this book will be concluded in the next chapter along with a discussion on current and future AM trends.

References

1. Kellens, K., Baumers, M., Gutowski, T. G., Flanagan, W., Lifset, R., Duflou, J. R., Environmental dimensions of additive manufacturing: mapping application domains and their environmental implications. *Journal of Industrial Ecology*, 2017, **21**(S1): pp. S49–S68.
2. National Institute on Aging, *Why population aging matters: a global perspective*, Publication No. 07–6134, 2007.
3. Huang, S. H., Liu, P., Mokasdar, A., Hou, L., Additive manufacturing and its societal impact: a literature review. *The International Journal of Advanced Manufacturing Technology*, 2013, **67**(5): pp. 1191–1203.
4. Rejeski, D., Huang, Y., *Environmental and health impacts of additive manufacturing: an NSF workshop report*, 2015, Woodrow Wilson Center, Washington, DC.
5. Rejeski, D., Zhao, F., Huang, Y., Research needs and recommendations on environmental implications of additive manufacturing. *Additive Manufacturing*, 2018, **19**: pp. 21–28.
6. Drizo, A., Pegna, J., Environmental impacts of rapid prototyping: an overview of research to date. *Rapid Prototyping Journal*, 2006, **12**(2): pp. 64–71.
7. Weller, C., Kleer, R., Piller, F. T., Economic implications of 3D printing: market structure models in light of additive manufacturing revisited. *International Journal of Production Economics*, 2015, **164**: pp. 43–56.
8. Peng, T., Kellens, K., Tang, R., Chen, C., Chen, G., Sustainability of additive manufacturing: an overview on its energy demand and environmental impact. *Additive Manufacturing*, 2018, **21**: pp. 694–704.

18

Conclusion

18.1 Conclusive Summary

This is the last chapter of Section D, which covers the trends, advancements, applications and conclusion. The previous chapter provided a detailed discussion on AM impact and forecasting. This chapter presents an overall conclusive summary to this book and a detailed discussion on the future trends of AM.

Additive manufacturing has a market niche with an enormous growth potential if the main barriers to up-take it can be addressed. While AM started mainly as a means to create prototypes, recent technological advancements and applications of AM technologies suggest that the technology has a potential to revolutionize many facets of everyday life. The last three decades have witnessed enormous technological and technical advancements in terms of the significance and contribution of AM systems. From the initial prototyping applications to present-day modelling, prototyping, tooling and manufacturing applications, AM technology is progressing leaps and bounds. AM has the capability to invoke the next industrial revolution and is predicted to globally foster economic contributions ranging from $200 billion to $600 billion in a year. Though a misnomer, AM is used interchangeably with 3D printing. Owing to the expiry of most of the primary patents in the early twenty-first century, huge industrial growth and advancements have been witnessed in the field of AM for improving its efficiency as well as effectiveness.

As is a common fact, AM is based on fabricating layered artefacts and has the unique potential of delivering highly customized parts. The need for supply chain management is totally eliminated, material wastage is tremendously reduced and manufacturing lead times are greatly shortened by the compression of the design cycle, thereby making it a subject of great interest amongst manufacturers. Since AM is a direct output-oriented manufacturing strategy, so great savings in energy and fuel consumption can be obtained thus reducing carbon footprints and greenhouse gases. Though AM was initially seen as a strategy to complement conventional manufacturing, today it has bypassed it in many applications.

One very important point to be kept in mind while dealing with AM options is that this field can experience growth in direct proportion to the adoption of an interdisciplinary approach. It will only then be possible to fully understand and solve the key research issues in the field of AM applications. Despite remarkable progress in the domain of AM, various aspects such as production speeds and quantities, precision, quality, materials, communication interfaces, etc. need attention to fully explore the potential of AM. Basic steps of component fabrication via the AM route are similar, so the quality of final product is mainly dependent upon the principle of building the component, and are also modeller specific. An efficient energy transfer mechanism capable of focussing and directing energy in order to melt/cure/soften material at the desired position is also a key requirement of any AM process and needs a lot of careful consideration.

Though AM is successfully utilized in various biomedical applications, challenges in the field of utilization of AM techniques, especially material related issues, need to be effectively overcome for their successful utilization in the development and application of biomaterials. Many research gaps lie in the field of standardizing AM related test samples, materials and parameters. Therefore, certification and benchmarking are also big AM challenges.

Hybrid AM processes that combine the unique benefits of additive with subtractive manufacturing methodologies offer numerous advantages and are a field of active research amongst AM personnel. Many AM limitations like restrictions on build volume, inability to fabricate multi-material parts, lesser transverse strengths, etc. can be easily overcome by the utilization of these hybrid systems.

18.2 Future Trends

This book summarizes the recent trends and numerous applications of AM in almost every field including tooling, aerospace, automobile, marine, construction, medical, retail and fashion, and so on. As has already been mentioned in many relevant sections, AM does not aim at the replacement of conventional manufacturing techniques. It is stretched and irrelevant to assert that it will revolutionize the manufacturing sector and render traditional set-ups obsolete. The fact is that it complements conventional manufacturing processes if its unique abilities are fully exploited and explored. AM is thus rarely seen as a standalone process, and the same trend is expected in the near future also. However, it has the ability to be effectively integrated into a multi-process system or multiple systems to keep pace with ongoing material advancements and newer product requirements.

AM has established itself with a firm footing in niche sectors like automotive, aerospace, biomedical and tissue engineering, etc. It has opened up entirely new manufacturing as well as supply chain opportunities. Betterment of existing fabrication opportunities and development of newer facilities that were initially thought impossible has become a reality with AM. Almost all industrial sectors ranging from industrial to retail have benefited from the utilization of AM technology and have been riding upon its various opportunities. It has opened up newer avenues with respect to manufacturing opportunities. It has substantially impacted the manner of product fabrication and the business strategies adopted by industrial players. It can be safely concluded that the manufacturing world has been presented with an ability to hop upon a decentralized industrial revolution and AM is one of most important key enablers of this trend.

Despite numerous advantages, many underlying limitations still need to be overcome to enable the utilization of these techniques for full-fledged manufacturing. Two main challenges in developing the next generation of AM processes include: (1) improvement in the speed and resolution of AM processes with lower energy consumption and (2) development of new 3D printing materials with localized control over mechanical, chemical and physical properties.

Apart from these, several challenges of CAD-based modelling need to be properly eliminated and overcome for effectiveness in DFAM techniques. A detailed understanding of relevant process parameters is required to completely understand the AM process. Since process parameters vary with the choice of AM process, so this exercise will take time and its generalization is quite difficult. This aspect will also play a crucial role in the selection of AM process for a particular application. A lot of facts and relevant information need understanding before choosing a particular AM process for any application, which in turn requires updated knowledge of the machines, materials, as well as benchmarking techniques. An overall improvement in the method of representation, user preferences and needs is also required. A better strategy for the generation, selection as well as evaluation of process chains, especially in the case of multi-component, multi-material and highly intricate parts, is required. Integration of relevant AM databases is also necessary. Optimization of AM processes in terms of build speeds, build times, build volumes, accuracy, dimensional tolerances, process parameters, mechanical properties and so on is required for all systems. Assistance tools to enhance process capabilities that possess the ability to predict part qualities need to be designed for effective utilization of these processes. Various underlying software issues need to be properly resolved. Generation of .stl files used with AM systems is a tedious process which is error prone and thus needs a lot of careful understanding.

Future aspects of AM are mainly based upon its capability to literally grow parts. Since DFAM is based upon functionality rather than manufacturing process, AM is much simpler, more effective and more efficient. It has opened up altogether new avenues of product customization, performance improvement, cost reduction, assembly line compaction as well as product conceptualization. It is expected that the mechanism by which a product is conceived and developed will be revolutionized in years to come. End user-based customization of highly intricate products in smaller as well as larger numbers and the WYSIWYB ability is thus completely feasible. This aspect of AM has revolutionized the entire industry and has widespread applications in industries like biomedical, automotive, aircraft, etc. The AM industry has witnessed a lot of improvement in speed, quality, accuracy and material properties over recent years. It will thus not be an exaggeration to summarize that AM systems possess the ability to be utilized as home fabrication devices.

The concept of "digiproneurship," which is based upon advanced information, communication as well as manufacturing technologies, has emerged with advancements in AM technology. This leads to a remarkable reduction in non-value addition resources, as well as activities like warehousing, infrastructure and tooling needs, etc. during fabrication. This technology can help in the reduction of regional limitations to a large extent. The unique capabilities of AM are bound to make definite technical, societal as well as economic changes. Advent of newer and advanced AM technology is making it an extremely promising manufacturing option and has hence enhanced its market share. Minimal design restrictions, non-existent tooling requirements, manufacturing versatility, customized raw material in the form of functionally graded materials or multi-materials, etc., ease of integration with conventional manufacturing techniques, ability to fabricate objects with virtually any number of subparts or features, ability to manufacture parts of any intricacy, etc. make AM an unparalleled manufacturing paradigm.

A number of textbooks/edited books on AM are available. However, the present work fully addresses the syllabus requirement of various leading and progressive universities which have incorporated AM into the curriculum at various educational levels. This book is basically an attempt to discuss the known AM concepts for learners and introduce the future aspects to professionals/researchers. Aspects related to various types of AM techniques, materials utilized, applications, trends, etc, are discussed in detail in this book. Various challenges and advancements are discussed throughout to help researchers in recognizing the probable areas of innovation and research. Different AM aspects have been exhaustively covered in this book. Being a part of university course designing process, the authors of this book fully understand the necessary aspects of AM that need to be mandatorily covered in a textbook to introduce readers to basic as well as advanced aspects. This book is divided into four sections to facilitate the orderly arrangement of different aspects. Though this book is written in an extremely simplified fashion, it is expected that readers have a primary understanding of CAD, networking, computer applications, materials, manufacturing principles, mechatronics, etc.

As a conclusion to this book, it can be said that AM is a change in manufacturing mindset that has the capability to both substitute as well as complement conventional manufacturing. It is the judiciousness of choice and decision of the manufacturing practitioner and engineers to come up with the optimum integration of both these manufacturing methodologies. AM is at the crossroads of its transition from a newer and much-hyped yet somewhat unproven manufacturing process towards a direction where it exhibits an ability to fabricate real, innovative, complex and robust products.

This book is one modest initiative by the authors to share and come up with a few pearls in the ocean of AM. However, this account is not summative in the sense that much more lies ahead and is forthcoming even as this book reaches its readers. However, all efforts have been made to introduce the readers to the necessary aspects of AM, especially university level graduate and postgraduate students. The idea of drawing or visualizing something and possessing the ability to immediately transform it into reality is far more intriguing as well as fascinating in reality than it appears while reading about it.

Index

Page numbers in **bold** denote tables, those in *italics* denote figures.